Alteration of Ovoproducts

Series Editors
Jack Legrand and Gilles Trystram

Alteration of Ovoproducts

From Metabolomics to Online Control

Edited by

Olivier Gonçalves
Jack Legrand

ISTE Press
ELSEVIER

First published 2018 in Great Britain and the United States by ISTE Press Ltd and Elsevier Ltd

ISTE Press Ltd
27-37 St George's Road
London SW19 4EU
UK

www.iste.co.uk

Elsevier Ltd
The Boulevard, Langford Lane
Kidlington, Oxford, OX5 1GB
UK

www.elsevier.com

British Library Cataloguing-in-Publication Data
A CIP record for this book is available from the British Library
Library of Congress Cataloging in Publication Data
A catalog record for this book is available from the Library of Congress
ISBN 978-1-78548-271-7

Contents

Acknowledgements

PÔLE
AGRONOMIQUE
OUEST

Our warmest thanks go to Mr. Henry Freulon of the Pôle Agronomique Ouest (PAO) without whom this initiative would never have seen the light of day. The PAO is a tool created by the French regions of Brittany and Pays de la Loire in 1992 to initiate and motivate research programs searching for solutions to economic challenges in the area. Its action relies, among other things, on groups made up of companies from the same sector, thus competing with each other, who agree on their expectations and then participate in the monitoring of multidisciplinary research programs.

The projects thus created and piloted, on the one hand, increase key knowledge for western economic sectors, and on the other hand, structure collaborative networks between companies, unions, technical centers, schools and research teams in the area.

For more information: https://www.pole-agro-ouest.eu

ADRO Ouest

Association pour le **D**éveloppement de la **R**echerche sur les **O**voproduits dans l'**Ouest**.

We would also like to thank Mr. Olivier Galet, president of ADRO Ouest, for his support and trust. It was through the initiative of the Pôle Agronomique Ouest that egg product companies were asked in 1994 to participate in a reflection of their future needs for new knowledge. Indeed, France, with the regions of Brittany and Pays de la Loire as its main production area, is the leading European producer of eggs and egg products, ahead of Germany and Italy. To date, about 40% of shell eggs are processed into egg products (a proportion that is steadily increasing). A dynamic is then immediately created, initially with small applications projects to show the interest of this type of collaboration between competitors (research and pre-competitive development). Project after project, the number of research organizations studying egg products increased and the group was structured into an association law 1901 named "Association for the Development of Research on Egg Products in France" (ADRO Ouest) in 2006.

To date, 38 collaborative programs have been piloted, including four theses and two ANR programs.

Introduction

An egg product is a complex matrix composed of proteins, lipoproteins, lipids and micronutrients. In its shell egg form, the egg product is particularly well protected from spoilage; the mechanical properties of the shell and its membranes, and bacteriostatic properties associated with the egg white keep the egg for several weeks at room temperature without spoiling. Once broken and homogenized, the whole liquid egg product loses its conservation properties and becomes an excellent growth medium for some bacteria. The bacterial activity will then cause spoilage of the liquid egg product matrix due to the consumption of endogenous metabolites and the production of exogenous metabolites. The chemical signatures of the liquid egg product will then evolve with incubation time, eventually causing a change in its quality. Clarification of the overall chemical composition of liquid egg products is therefore a major challenge for the maintenance of its quality and the monitoring of its spoilage during production processes. To date, the early detection of chemical changes of batches of egg products during their production processes or conservation is not possible.

It is in this context that this book proposes to make an inventory of the latest analytical approaches that could be used to control the spoilage of liquid egg products during their production processes. The complexity of the egg-product matrix will be detailed initially (Chapter 1), with the potential sources of biotic and abiotic spoilage. Known biomarkers will also be discussed (Chapter 2). Then the exploratory possibilities of the latest analytical techniques resulting from metabolomics will be exposed. Invasive approaches (chromatography coupled or not with mass spectrometry) and non-invasive approaches (magnetic resonance or vibrational spectrometry) will be more specifically detailed (Chapter 3). Finally, the latest generation of sensors capable of measuring the process as closely as possible (in-line) will be presented as well as the transferable techniques derived from high-resolution analytical approaches. The potential of these approaches will also be discussed as well as their application as control techniques for the spoilage of liquid egg products (Chapter 4).

1

Presentation of the Ovoproduct Matrix

1.1. Structure and composition of hen eggs

Due to intensive selection for many years and the genetic homogeneity of the strains of laying hens currently used, the average weight of a chicken egg varies little, ranging from 55 to 65 g. The egg consists of three main parts: the shell (about 10% of the weight of the egg), white or albumen (60%) and yolk or vitellus (30%).

The shell and associated shell membranes, although inedible, are essential elements for the quality of the egg contents because of their role as a physical barrier to microbiological contamination. The shell itself is essentially mineral in nature (95% (w/w) mineral, of which 93.5% is calcium carbonate), while the cuticle covering is organic in nature, as are the two shell membranes. These membranes, which separate the shell from egg white, are constituted by the superposition of layers of crossed protein fibers. They are a very effective barrier to bacteria and molds which may possibly cross the pores of the shell when the cuticle is damaged.

Chapter written by Marc ANTON, Françoise NAU, Valérie LECHEVALIER and Anne MEYNIER.

Egg white is principally an aqueous solution of proteins, carbohydrates and minerals. See Table 1.1. Despite this relatively simple overall composition, it is a heterogeneous medium which is divided into four distinct layers in the freshly laid egg:

– the chalaziferous layer, which is very firm and surrounds the vitelline membrane, and extends towards the two extremities of the egg by the chalazes. It represents 3% (w/w) of the total albumen;

– the external liquid albumen (23%) in contact with the shell membranes;

– the thick albumen (57%) fixed at both ends of the egg and having a gelled structure;

– the inner liquid albumen (17%) surrounding the yolk.

These different fractions do not have the same water content (from 84 to 89%, from the inner layers to the outer layers of the egg), nor exactly the same protein concentrations. The thick egg albumen would thus be four times more concentrated in ovomucin than the liquid egg albumen, giving it a gelled structure and a much higher viscosity. In addition, physicochemical changes occur during the storage of the eggs. These changes essentially consist in a pH increase (from 7.5 at the time of laying to 9.5 after a few days) due to CO_2 loss, resulting in liquefaction of egg albumen and transformation of ovalbumin into more thermostable S-ovalbumin.

	Whole egg	Albumen	Yolk
Water	76	88	50
Proteins	12.5	10.6	16
Lipids	10.5	–	33
Carbohydrates	0.5	0.8	0.5
Minerals	0.5	0.6	0.5

Table 1.1. *Mean composition of whole egg, albumen and yolk, expressed as % of total weight [THA 94]*

Egg yolk has a solids content close to 50% (see Table 1.1) consisting mainly of lipids (about 65% of dry matter) and proteins (33%). It also contains a significant amount of minerals including calcium, iron and phosphorus. Egg yolk is a dispersion of particles (profiles and granules) in equilibrium in an aqueous solution of proteins (see Figure 1.1).

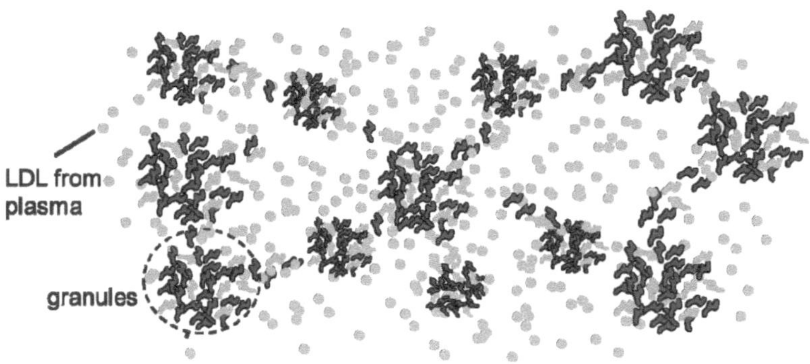

Figure 1.1. *Representation of the internal structures of egg yolk [ANT 13]. For a color version of this figure, see www.iste.co.uk/goncalves/ovoproducts.zip*

The profiles are assembled structures from 12 nm to 48 nm in diameter and are considered as low-density lipoproteins (LDL). The granules have the shape of more or less flattened spheres with a diameter of 0.3 µm to 2 µm. By diluting in NaCl (0.17M) and centrifuging (10000g), the egg yolk can be separated into two fractions:

– plasma (orange supernatant) accounts for 75–80% of the yolk dry matter and includes LDL (85%) and 15% of soluble proteins named livetins. It contains about 55% of proteins and 85% of phospholipids of yolk;

– granules (pale pellets) constitute 20–25% of the yolk dry matter and contain high-density lipoprotein (HDL) (70%), phosvitin (16%) and residual LDLg (12%) (see Table 1.2). The granules represent about 47% of the proteins and 15% of the phospholipids of yolk. HDL and phosvitin, forming the basic link of the granules, contain a high proportion of phosphorylated serine residues which allow for their association via divalent calcium ions. The numerous phosphocalcic bridges make the structure compact, slightly hydrated, poorly accessible to enzymes and protect the proteins against denaturation and thermal gelling.

		% dry matter of yolk	% lipids of yolk	% proteins of yolk	% lipids	% proteins
Yolk		100	100	100	64	32
Plasma		78	93	53	73	25
	LDL	66	61	22	88	10
	livetins	10	-	30	-	96
	others	2	-	1	-	90
Granules		22	7	47	31	64
	HDL	16	6	35	24	75
	phosvitin	4	-	11	-	95
	LDLg	2	1	1	88	10

Table 1.2. *Repartition of yolk constituents [POW 86]*

Finally, the whole egg is constituted of 76% water and the dry matter is distributed in quasi-equivalent proportion between proteins and lipids. This composition makes the egg a low energy food (6.25 Jg^{-1}). It is also characterized by the very high nutritional quality of its proteins (absence of limiting amino acid for the adult) and its lipids (very good digestibility with a natural abundance in unsaturated fatty acids), as well as its richness in phosphorus, iron and many vitamins. This universally consumed food is however low in calcium, totally devoid of vitamin C and fiber, and constitutes a major food allergen, especially in young children.

1.2. Biochemical and structural characteristics of egg proteins and lipids

1.2.1. *Albumen proteins*

Proteins constitute most of the dry matter of egg albumen with a ratio of nitrogenous matter/solid content higher than 90%. The total number of proteins in egg white is not precisely known. Until very recently, only major proteins were identified. However, the recent development of powerful separative and analytical techniques has revealed many minor proteins, some of which have already been identified. The main characteristics of the currently known proteins are shown in Table 1.3.

Protein	%	Experimental MW (kDa)	Experimental pH_i	Important characteristics
Ovalbumin	54	45	4.5	Phosphoprotein, immunogenic
Y Ovalbumin	7	53.6	5.4	nd
X Ovalbumin	0.5	53	6.0-6.7	nd
Ovotransferri n	13	77.7	6.1	Iron chelator, bacteriostatic activity
Ovomucoïd	11	28	4.1	Trypsin inhibitor
Ovomucin	1.5-3.5	5600 to 23000	4.5-5.0	Glycosylated, viral hemagglutination inhibitor
Lysozyme	3.5	14.3	10.7	Lyse the walls of the Gram + bacteria
Ovoinhibitor	0.1-1.5	65	6.2-6.4	Serine protease inhibitor
Ovoglyco-protein	0.5-1	40	5.2	nd
Flavo-protein	0.8-1	29.2	4.0	Riboflavine (vitamin B2) chelator
Ovostatin	0.5	780	4.9	Serine protease inhibitor, allergenic

Protein	%	Experimental MW (kDa)	Experimental pH_i	Important characteristics
Cystatin	0.05	12.7	5.1	Cysteine-protease inhibitor
Avidin	0.05	68.3	10	Biotine (vitamine B8) chelator
Ex-FABP (Ch21)	nd	21	5.7	Lipocalin family
CAL gamma	nd	22	5.6-6.0	Lipocalin family
TENP	nd	48.9	5.9-6.3	Bactericidal permeability-increasing protein family
Hep 21	nd	18	6.4	uPar/Ly6/Snake neurotoxin family

nd: not determined

Table 1.3. *Protein composition of albumen. In italic are indicated minor proteins recently identified [LI 89, GUE 10]*

Proteins of egg albumen are predominantly globular proteins (except ovomucin) with an acidic isoelectric point (except lysozyme and avidin), and are all glycoproteins (except the major form of lysozyme as well as cystatin). They are also rich in sulfur amino acids. Many of these proteins possess biological properties, especially antibacterial properties contributing to the protective role of egg white during development of the embryo. Some of them are highly thermosensitive and/or sensitive to surface denaturation, thus contributing to the remarkable techno-functional properties of egg white.

Ovalbumin, the major egg white protein, alone accounts for more than half of the total proteins. It is a globular protein, phosphorylated (mainly 2 phosphates per molecule), with a molecular weight of about 45 kDa. It belongs to the serpin family although it lacks protease inhibitory activity. Among the 385 amino acid residues constituting the protein, half are hydrophobic and one third are charged, mostly negatively at physiological pH. Ovalbumin also has 6 Cys residues buried in the core of the protein; two of which are involved in a disulfide bridge (Cys73–Cys120). Thus, it is the only egg white protein that has free thiol groups which can be rearranged according to the conditions of preservation, pH and surface denaturation. During egg storage or after moderate heat treatments in an alkaline environment, ovalbumin acquires a more thermostable conformation called S-ovalbumin by isomerization of specific amino acid residues (Ser164, Ser236 and Ser320 which adopt a D-configuration).

Ovotransferrin, also known as conalbumin, belongs to the transferrin family. It is a polypeptide chain of 686 amino acid residues (about 77,7 kDa) organized into two lobes, each having a specific binding site for iron (or Cu^{2+}, Zn^{2+}, Al^{3+}). Ovotransferrin is the most thermosensitive protein of egg white: its denaturation temperature at pH 7 is about 63 °C making it the limiting factor of egg white pasteurization. The iron (or aluminum) binding to the active sites causes a change in the conformation of the protein and an increase in its thermal stability.

The ovomucoid is an acidic protein (pHi of 4.1) of about 28 kDa which can contain up to 25% carbohydrates (w/w). It is a major allergenic protein of egg white. At pH 7, its denaturation temperature is about 77 °C. On the other hand, it is very heat-resistant at acidic pH, since it conserves its antitrypsic activity after several minutes at 100 °C.

Ovomucin is also a highly glycosylated protein of high molecular weight (higher than 5,600 kDa). It is insoluble at low ionic strength, that is after egg white dilution in water, and after precipitation, its re-solubilization is extremely difficult. Electrostatic interactions can occur between ovomucin and other egg white proteins: ovalbumin, ovotransferrin and especially lysozyme. At the pH of the egg white (between 7.5 and 9.5), the carboxylic groups of the sialic acid of ovomucin interact with the ε–NH_{3+} groups of the lysyl residues of the lysozyme, forming a lysozyme-ovomucin complex which is insoluble in water, and would be responsible for the gelled structure of thick egg white. The dissociation of this complex at alkaline pH would be responsible for the egg white liquefaction observed during egg storage.

Lysozyme is an enzyme responsible for the lysis of the walls of Gram + bacteria by hydrolysis of the β-1,4 bonds between N-acetylmuraminic acid and N-acetylglucosamine of the bacterial wall peptidoglycans. It makes lysozyme the only egg protein purified and used at an industrial scale, for food and pharmaceutical applications. But other antibacterial mechanisms of lysozyme have recently been described, and other biological properties have also been described (antiviral, anti-inflammatory, immunomodulation). It is a globular protein consisting of 129 amino acid residues, about 40% of which are hydrophobic and one-third is charged. The lysozyme has a rigid structure stabilized in particular by four disulfide bridges (Cys6-Cys127, Cys30-Cys115, Cys64-Cys80, Cys76- Cys94).

1.2.2. *Yolk lipoproteins and proteins*

Low density lipoproteins (LDLs), present in the plasma, are the major constituents of egg yolk: they represent 2/3 of its dry matter and 22% of its proteins. They contain 83–89% of lipids and 11–17% of proteins. The lipids are distributed in 74% of neutral lipids (triglycerides and cholesterol) and 26% of phospholipids. Two types of LDL have been demonstrated: LDL1 (20% of LDL) with a molecular weight of 10.3×10^6 Da and LDL2 (80% of LDL) with a molecular weight of 3.3×10^6 Da.

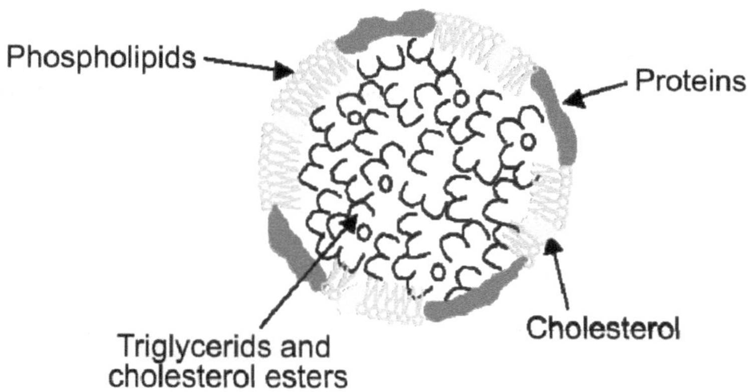

Figure 1.2. *Schematic representation of Low Density Lipoproteins from egg yolk. For a color version of this figure, see www.iste.co.uk/goncalves/ovoproducts.zip*

LDLs have a classic lipoprotein structure (Figure 1.2) with a core of neutral lipids (triglycerides and cholesterol esters) surrounded by a monolayer of phospholipids and proteins in contact with the aqueous phase. They have a spherical shape, their diameter varying from 17 to 60 nm. The LDLs of the egg yolk are derived from lipoproteins synthesized in the liver of the hen, the VLDL (Very Low Density Lipoproteins). The VLDLs are transported by the blood flow to the ovary where oocytes are present. Specific receptors in the oocyte membrane allow the fixation of the VLDLs which are then secreted in the egg yolk by endocytosis and then modified into LDLs.

LDL proteins are particularly difficult to study due to their low solubility and their high proportion of hydrophobic amino acids (40%), which places the LDL among the most hydrophobic proteins existing. Moreover, they are glycosylated and have a pHi between 6.5 and 7.3.

Livetins are non-lipid globular proteins and are present in the plasma. These proteins represent 11% of the dry matter of the yolk and 30% of its proteins; their pHi is between 4.3 and 5.5. They are blood proteins deposited in the yolk.

High density lipoproteins (HDLs), present in the granules, account for about 1/6 of egg yolk dry matter and 36% of its proteins. They are made up of two subunits α and β whose amino acid composition is very close but which differ by their phosphorylation degree and their sialic acid content. They contain 80% proteins and 20% lipids, which are divided into 65% phospholipids and 35% neutral lipids.

Phosvitin, present in the granules, accounts for 11% of yolk proteins and 23% of granules proteins. The molecular weight of phosvitin is between 36 and 40 kDa and its pHi is 4.0. It is glycosylated and contains about 10% phosphorus (60% of yolk phosphorus). More than half (54%) of the amino acid residues in phosvitin are serine, which are exclusively present as phosphoric acid esters, conferring to the protein high cation-binding properties, mainly for iron and calcium. Moreover, the polypeptide sequence of phosvitin contains little or no cysteine, cystine, methionine, tryptophan and tyrosine residues, and only 10% hydrophobic amino acid residues. Phosvitin is, therefore, particularly hydrophilic in comparison with most proteins and carries a strongly net negative charge at physiological pH conferring to it a rod-shaped structure.

1.2.3. *Yolk lipids*

The main components of yolk (60%), lipids are distributed exclusively in lipoproteins (LDLs and HDLs). They are composed of triglycerides (65%), phospholipids (29%), cholesterol (5%), free fatty acids (<1%) and other lipids including carotenoids (<0.1%). The cholesterol level derived from egg is lesser than the cholesterol concentration synthesized by humans [SIN 12]. Yolk phospholipids are very rich in phosphatidylcholine (76% phospholipids) and phosphatidylethanolamine (22%).

The fatty acid composition of egg lipids, based on standardized feeding of hens, is about 30–35% saturated fatty acids, 40–45% monounsaturated fatty acids and 20–25% polyunsaturated fatty acids. The main fatty acids are oleic acid (C18:1 to 40–45%), palmitic acid (C16:0 to 20–25%) and linoleic acid (C18:2 to 15–20%). However, this composition is subject to large variations, in particular depending on the nature of dietary fatty acids of the hen.

The health benefits of n-3 fatty acids are nowadays widely recognized, but their intake is below the recommendations in most western countries. Eggs are an interesting target, as they are part of a regular western diet, and their FA composition can be modified through the diet of the hen. The enrichment of the diet can be achieved by three main sources: vegetable oil (mainly flaxseed oil or extruded flaxseed), fish oil or recently algae oils as an alternative to fish oil. Vegetable oils provided α-linolenic acid (18:3 n3), whereas fish and algae oils provided long chain n-3 FA such as 20:5 n-3 (EPA) and 22:6 n-3 (DHA). Since the late 1990s, n–3 PUFA enriched eggs produced through feed supplementation with flaxseed have become available on the market in many countries [SUR 01]. The level and source of omega-3 fatty acids (flaxseed oil, algal DHA) have no impact on hen performance and shell quality [NEI 16], nor on the lipid and cholesterol content of egg yolk. The addition of fish oil to the hens' diet gave contradicting results on hens' production parameters. Several studies highlighted a decrease of egg and/or yolk weight upon fish oil feeding, whereas other studies mentioned no impact on the egg weight [FRA 12].

From a recent review [FRA 12], it can be concluded that when flaxseed is added to the hen's diet the increase of the level of ALA in yolk lipids was proportional to that of the diet. The level of DHA increased as well, but to a lesser extent and it was not in linear response to the level of flaxseed. This is the result of the poor yield of the elongation and desaturation pathways in yolk as in human metabolism that converts ALA into EPA and DHA.

Enrichment of hens' diet with n-3 FA results in an increase in the proportion of n-3 in egg yolk lipids. Nevertheless, this change in fatty acid composition affects TAGs and PLs differently. Thus, the proportions of ALA increase in triacylglycerols (TAGs) when the diet contains vegetable oil such as rapeseed or flaxseed oil, whereas the proportions of long chain PUFA such as DHA increase in the phospholipids [FRA 12, MEY 14]. The same trends are observed in the case of the addition of fish oils or microalgae to the feeding of hens. These results are important in regard to the bioavailability of the FA being higher from PL than from TAG.

As the quantities of n-3 PUFA that can be deposited in eggs are restricted by hens' physiology, some researchers proposed an alternative technological approach making it possible to simultaneously enrich eggs products with n-3-PUFA and reduced cholesterol [KAS 10, TAM 17].

1.3. Physico-chemical properties and applications

1.3.1. *Interfacial properties*

1.3.1.1. *Foaming properties of albumen proteins*

Egg albumen is the perfect foaming agent: compared to other plant or animal protein ingredients, it offers higher foaming properties. These properties are due, on the one hand, to the very good interfacial properties of the albumen proteins, and on the other hand, to its ability to fix this expanded structure, especially after heat treatment, thanks to many interactions that can be established between proteins, inside the interfacial layer and between the latter and the underlying proteins. Actually, the ability to overrun the barrier of globular proteins such as the egg white proteins depends on the implementation of 3 phases occurring on the surface or in the vicinity of the gas bubbles:

– diffusion of proteins to the air-water interface;

– conformational changes and compaction of adsorbed proteins at the interface;

– irreversible rearrangement of the protein film.

This is why the foaming ability of egg white depends on intrinsic factors, including the structure and conformation of the proteins, themselves dependent on environmental factors such as pH, ionic strength, protein-protein and protein-water interactions.

Overall, the egg albumen can be assimilated to a solution of effective surfactants. Its proteins have a relatively large surface hydrophobicity, thus diffusing rapidly towards the air-water interface and adsorbing efficiently. Their molecular flexibility allows for conformational rearrangements (expansion) at the interface that causes a significant reduction in surface tension. Their ability to form an intermolecular continuous network, in particular when a certain level of denaturation has been previously achieved, enables them to constitute a visco-elastic interfacial film ensuring the stability of the foam. However, the egg white proteins do not all exhibit the same characteristics to the same extent, and therefore do not all participate equally in the foaming properties. Particularly, globulins have long been recognized for their foaming power, whereas ovomucin would confer foam stability at ambient temperature, and ovalbumin would permit foam structuring after heating.

But it is now well established that the complexity and the synergy of the phenomena do not make it possible to dissociate the roles of the different egg white proteins. Thus, ovalbumin, lysozyme and ovotransferrin exhibit different behaviors at the air-water interface depending on whether they are alone in solution or in a mixture [LEC 03].

Parameters such as surface hydrophobicity (which determines the efficiency for interfacial adsorption), the number of disulfide bridges (which determines the rigidity/flexibility of the protein) and the number of free thiol groups (which determines the "reactivity") are decisive in the structural modifications taking place at the air-water interface.

Competition phenomena for the interface and possible exchanges between proteins at the interface can take place over time. In a mixture of ovalbumin, globulins, ovotransferrin, lysozyme and ovomucoid in the natural proportions of the egg white, [DAM 98] have shown that only ovalbumin and globulins adsorb at the interface. The exclusion of ovotransferrin, lysozyme and ovomucoid would be related to their lower diffusion rates: arriving at the interface after the two other proteins, they would not be able to move them. But in a simpler mixture of ovalbumin, ovotransferrin and lysozyme in the natural proportions of egg white, [LEC 05] were able to demonstrate the participation of the three proteins in the formation of the interfacial film, in which electrostatic interactions and covalent bonds were observed. Moreover, the analysis of the structural modifications undergone by each of these three proteins after contact with the air-water interface has revealed a phenomenon of "denaturation synergy": whether it is ovalbumin, ovotransferrin or lysozyme, the structural modifications following contact with the interface are higher or equal when these proteins are in a mixture with respect to the modifications observed when each of the proteins is alone in solution.

That said, a number of studies have shown that the foaming properties of isolated egg white proteins are still much lower than those of egg white, which tends to confirm the existence and the role of protein interactions. In particular, it is commonly accepted that the natural coexistence in egg albumen of alkaline proteins (lysozyme) and acidic proteins (ovalbumin) is responsible for electrostatic interactions, partly explaining the good stability of albumen foams.

1.3.1.2. *Emulsifying properties of yolk lipoproteins*

Egg yolk-based emulsions are oil-in-water emulsions such as mayonnaise, bearnaise, custard, dressings, etc. that is to say they consist of oil droplets dispersed in an aqueous phase. An emulsion is a thermodynamically unstable system because of the high interfacial tension created at the interface between the two immiscible liquids: the greater the interfacial tension, the more the emulsion will tend to destabilize rapidly. The main forms of destabilization are: creaming, flocculation and coalescence. All droplet aggregation states such as flocculation and creaming, or the use of large oil fractions promote coalescence. However, it is possible to form kinetically stable emulsions by slowing the aggregation or migration of the droplets; this can be obtained by adding emulsifiers and/or thickeners before homogenization.

To understand which is the main contributor of the emulsifying properties of egg yolk, many authors have separated yolk into its main fractions: plasma and granules. By comparing the stability and granulometry of yolk, plasma and granules emulsions, [DYE 93] and [ANT 97] found similarities between yolk and plasma emulsions, while granules emulsions showed markedly different properties (Figure 1.3) depending on their solubility [ANT 00]. Under emulsification conditions resulting in oil droplets smaller than one micrometer, the droplet size of emulsions prepared from insoluble granules is higher due to greater coalescence. On the other hand, when the granules are soluble, coalescence is avoided, and the oil droplets have a size comparable to those obtained with egg yolk. However, at the same protein concentration, granules remain less effective than plasma. This suggests that the molecules responsible for the emulsifying properties of egg yolk are in the plasma.

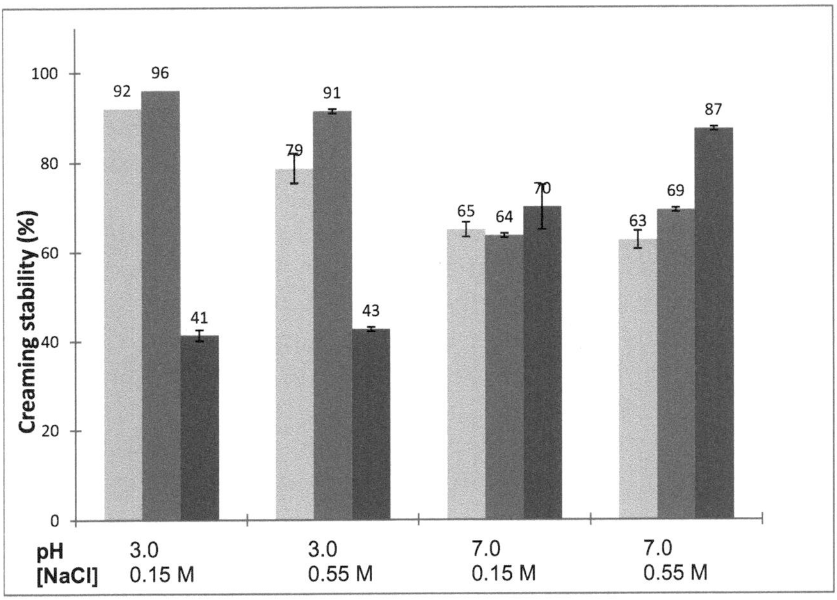

Figure 1.3. *Creaming stability of oil-in-water emulsions stabilized by yolk (yellow), plasma (orange), and granules (blue), Oil volume fraction: 30%, 25 mg protein/ml [ANT 00]. For a color version of this figure, see www.iste.co.uk/goncalves/ovoproducts.zip*

Among the constituents of plasma, many authors have shown that LDLs are better emulsifiers than bovine serum albumin (BSA) [MIZ 05] and β-casein [SHE 79]. These authors and others [ALU 98, MIN 00] have shown that LDLs rapidly displace caseins from the interface, demonstrating their excellent surface properties. The soluble yolk proteins do not have the penetrating power of lipoproteins at the oil-water interface, especially at high surface pressures. But, with equivalent dry matter, they form finer and more stable emulsions with respect to creaming. Conversely, livetins fail to adsorb when emulsified in a mixture with casein or with egg yolk LDLs. These results demonstrate the primordial role of LDLs, among the constituents of plasma, and hence of yolk. Moreover, when comparing the emulsifying properties of LDLs and HDLs in egg yolk, [MAR 02] showed that LDLs result in smaller oil droplet sizes than those obtained with HDLs, regardless of pH and ionic strength conditions. The synthesis of all these results indicates that LDLs play the major role in the high emulsifying properties of egg yolk.

It is usually assumed that the LDL micelles break in the vicinity of the interface by weakening protein-protein interactions. Then, the lipids of the LDL core coalesce with the oil phase of the emulsion, and the proteins and phospholipids spread out at the interface to surround the oil droplets [SHE 79, KIO 89]. In comparison, direct adsorption of the proteins and phospholipids that make up the outer LDL film is not so easy because of their low water solubility (high hydrophobicity). Thus, the interactions between proteins and phospholipids which contribute to the native LDL structure are essential to transport these non-soluble surfactants in a pseudo-"soluble" form in the vicinity of an interface and then to release them at the interface.

Langmuir's study of the LDL compression isotherm in comparison with those obtained with their constitutive molecular species (proteins, neutral lipids and phospholipids) made it possible to better understand the adsorption mechanism of LDLs at the air-water interface [MAR 03]. Three phase transitions at 20, 41 and 54 $mN.m^{-1}$ are observable on the LDL compression isotherm (Figure 1.4). The single transition observed on the neutral lipid isotherm corresponds exactly to the visible transition on the LDL isotherm at 20 mN.m^{-1}. The first transition of the LDL film is thus attributed to the neutral lipids which are therefore well present in the interfacial layer formed by the LDLs and undergo a rearrangement within the film at the pressure of 20 mN.m^{-1}. Neutral lipids form the core of lipoproteins and their presence in the LDL film necessarily involves the dissociation of LDLs at the air-water interface.

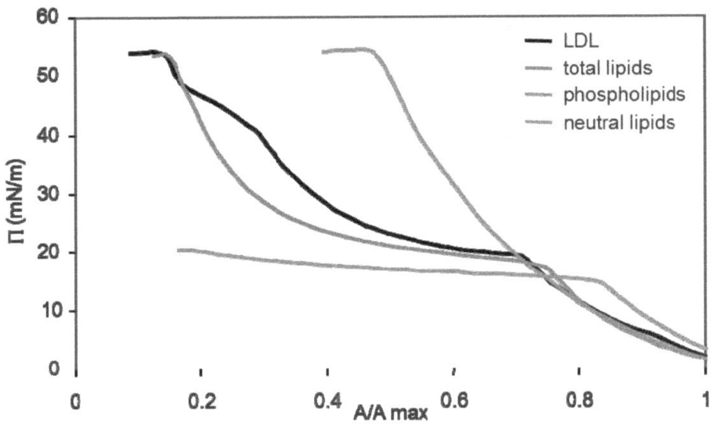

Figure 1.4. *Compression isotherm at the air-water interface of LDL, and various lipid constituents extracted from LDL [ANT 13]. For a color version of this figure, see www.iste.co.uk/goncalves/ovoproducts.zip*

The LDLs must therefore be dissociated after contact with the interface to allow for the release and spreading of the neutral lipids. Moreover, the observed transition on the phospholipid isotherm (54 mN.m^{-1}) corresponds exactly to the last visible transition on the LDL isotherm. The total lipids isotherm also exhibits this characteristic transition at 54 mN.m^{-1}, which is therefore attributed to the interfacial restructuring of the phospholipids. On the other hand, given their insolubility, the LDL proteins could not be spread out. The transition obtained at 41 mN.m^{-1} could not, therefore, be assigned, although the presence of proteins is widely suspected.

Given these results, it therefore seems obvious that the LDLs should break on contact with the interface and spread out on the surface, thus allowing the adsorption of their constituents. LDLs can therefore be considered as a vector for emulsifying agents (proteins and phospholipids) that could not be soluble in water and adsorbed once they are released near the interface.

The comparison of phospholipid isotherms in solution or in the form of vesicles (liposomes) with those of LDLs was carried out in order to understand the respective roles of proteins and phospholipids in the adsorption of LDLs. The phospholipids in the form of liposomes showed a low surface pressure throughout the isotherm. This means that the liposomes do not have the ability to destruct and spread to the interface. Conversely, proteins on the surface of the liposomes make it possible to obtain isotherms resembling those of LDLs. It was thus evidenced that the proteins located on the surface of the LDLs have an initial role of anchoring at the interface [ANT 13]. This anchoring causes protein denaturation which leads to the destabilization of the LDL outer layer and to the total spreading of the LDLs at the interface.

Atomic force microscopy showed that the second transition of the LDL compression isotherm was due to the demixing of protein-triglyceride complexes, whereas proteins and phospholipids appeared to be miscible throughout the isotherm. This explains the interaction mechanism of species released during LDL adsorption at an interface (Figure 1.5). These experiments have been carried out, for practical reasons, at the air-water interface. At the oil-water interface (which is that of the emulsified media), triglycerides certainly merge with the oil phase and the interface in this case consists solely of protein-phospholipid interactions.

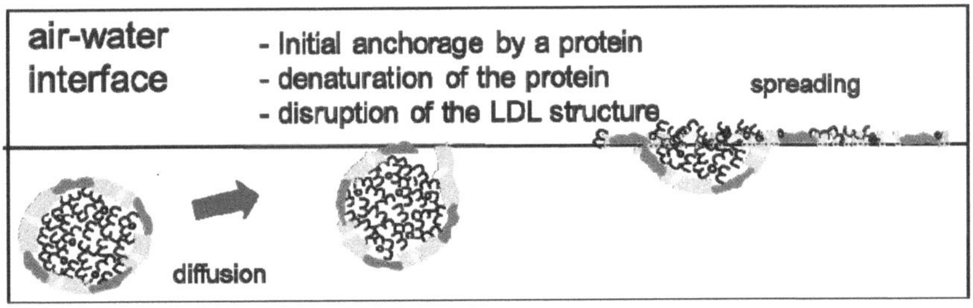

Figure 1.5. *Schematic description of the adsorption mechanism at the air-water interface of LDL of egg yolk and liposomes [ANT 13]. For a color version of this figure, see www.iste.co.uk/goncalves/ovoproducts.zip*

1.3.2. *Gelling properties*

1.3.2.1. *Albumen*

The gelling properties of egg albumen are used in many food applications using cooking methods (sausages, terrines, pastry, etc.). The physicochemical phenomenon exploited is the thermogelation of proteins, resulting in what is called "thermotropic protein gels".

Egg albumen loses its fluidity around 60 °C, but the maximum firmness is reached above 70 °C. Except ovomucin and ovomucoid, all egg albumen proteins coagulate when heated, despite different behaviors with respect to heat treatments. In particular, their denaturation temperatures strongly differ; for example, at pH 7 in egg white, they are around 85, 74 and 63 °C for ovalbumin, lysozyme and ovotransferrin, respectively [MAR 03]. Ovotransferrin is thus the most thermosensitive protein, which usually leads to considering it as the "initiator" protein of egg white gelation. Actually, the coagulation temperature of egg white and the rheological properties of the gels obtained can be strongly increased by selective elimination of ovotransferrin, or by metal ion (Al^{3+}, Fe^{3+}) chelation which increases the protein thermostability.

As a consequence of a change in the balance of attractive and repulsive forces between proteins, the thermogelation of egg albumen proteins perfectly corresponds to the model of the thermo-induced gelation of globular proteins. This is basically a two-stage phenomenon: thermal denaturation of proteins (first step), followed by aggregation of denatured proteins (second step).

The macroscopic characteristics of the gels obtained (firmness, breaking strength, elasticity, water retention capacity, etc.) depend on the number and nature (energy) of the interactions created, as well as on the characteristics of the gel network structure (particulate gel vs. filamentous gel). In thermotropic gels, the interactions involved are essentially hydrophobic and electrostatic (low energy interactions), but covalent bonds (disulfide bonds) of high energy also exist. Regardless of the type of interactions, their implementation depends on the conformation of the proteins, that is the level of denaturation reached at the end of the first step, since it determines the exposure of reactive regions. The interaction implementation also depends on the environmental conditions (mainly pH and ionic strength) which are able either to favor or to limit them, accelerating or slowing down the aggregation step. Thus, depending on the protein, and depending on the environmental conditions, protein aggregation occurs on more or less denatured proteins, leading to different gel microstructures. Therefore, the ratio between the speed constants of the two steps of egg white protein thermogelation is a determining factor.

These mechanisms have been particularly studied and demonstrated in the case of ovalbumin thermogelation. In particluar, the impact of ionic strength conditions on the structure and characteristics of the gels obtained has been extensively investigated. Thermal denaturation results in an overall increase in the surface hydrophobicity of ovalbumin. When the heating is applied at high ionic strength, the surface charges of the protein are screened, resulting in a decrease in electrostatic repulsions and thus promoting hydrophobic interactions. Under these conditions, random aggregates of weakly denatured proteins rapidly appear, leading to opaque gel, with lower elasticity, stiffness and water retention capacity. On the contrary, at low ionic strength (and at pH far away from ovalbumin pI), electrostatic repulsions are such that, on the one hand, they delay aggregation, which is favorable to a greater denaturation (predominance of the first step), and on the other hand, this aggregation can only occur through specific contact points (hydrophobic zones), which results in the formation of linear aggregates (Figure 1.6). After extended heating, these linear aggregates finally interact to form translucent gels, with higher elasticity, firmness and water retention capacity.

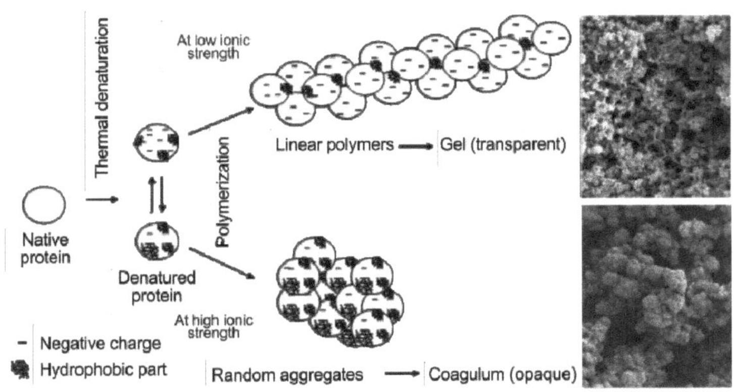

Figure 1.6. *Impact of ionic strength on ovalbumin thermogelation: mechanistic model [DOI 93] and observations by electron microscopy of the corresponding egg white gels [CRO 02]*

Ovalbumin is the major egg white protein, and thus plays a major role in the gelling properties of albumen. The impact of ionic strength (and pH) as described can then be transposed to egg white; a two-step heating process has been proposed to prepare egg white transparent gels, very firm and elastic, with an exceptional water retention capacity [KIT 88]. The first step consists of heating egg albumen previously desalted by dialysis in order to create soluble linear polymers made of extensively denatured proteins. And the second step consists of a heat treatment performed after salt addition, in order to enable hydrophobic interactions between the linear polymers previously created. In the same way that the ionic strength acts through the modification of the apparent charge of the proteins, pH is also a major parameter influencing egg white gelation. Near their pHi, proteins tend to form random aggregates, similar to what has been described at high ionic strength, because of low electrostatic repulsions. Thus, minimal gelling properties of egg white are obtained at about pH 5 (close to ovalbumin pHi), while the best gelling properties are obtained at alkalin pH (around 9). In these latter conditions, strong electrostatic repulsions delay aggregation, enabling increased denaturation, in favor of linear aggregates. Moreover, the greater reactivity of the thiol groups in this pH range also probably contributes to these improved gelling properties, due to easier disulphide exchanges.

1.3.2.2. *Yolk*

Gelation occurs when egg yolk is subjected to a heat treatment or a freeze-thaw process. With a dry matter content of 24%, a NaCl concentration of 0.17 M and a pH of 6.1, egg yolk and plasma gel after 2.5 minutes at 72 °C, whereas granules do not gel (Figure 1.7). The constituents responsible for the gelation of egg yolk are therefore plasma constituents. This particularly concerns LDLs whereas other yolk constituents do not or very poorly participate.

Gelation of plasma solutions (and therefore of egg yolk) is favored at neutral pH and/or high concentration of NaCl. These conditions correspond to a strong neutralization of the

LDL protein charges, which limits the electrostatic repulsion intensity at short distance, and thus promotes aggregation between proteins. The gelation times recorded, regardless of the pH and ionic strength conditions, are very short. This is specific to coagulation processes. Gels are opaque, a sign of light diffraction by high-sized particles (protein aggregates).

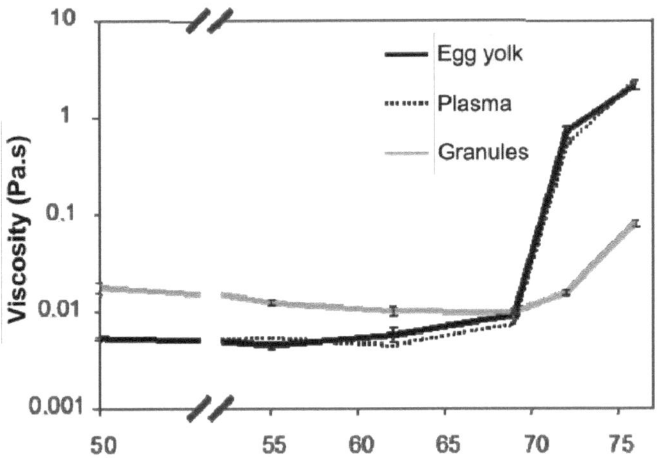

Figure 1.7. *Effect of temperature on the viscosity of egg yolk solutions, plasma and granules, pH 6.1, 0.17 M NaCl, 24% dry matter [LE 99]*

The apparent viscosity of yolk and plasma after heat treatment is strongly correlated with loss of protein solubility. The amount of soluble proteins decreases when gelation takes place. On the contrary, apparent viscosity and protein solubility are poorly correlated for the granules.

At ionic strength 0.17 M and pH 6.1, the granules seem to be able to withstand higher thermal treatments than yolk and plasma without undergoing gelation. The structure of granules is preserved under the imposed conditions of the medium. In these conditions, constituents of granules, in particular phosvitin and HDLs, are kept in aggregates. Although the granular structure does not prevent the denaturation of some of the proteins (HDL-α and LDLs), the resistance of phosvitin and HDL-β allows the granules to retain their structure, limit their aggregation and does not permit formation of a gelled network in the entire volume.

On the contrary, after previous solubilization of the granule constituents (0.5 M NaCl), gelation occurs during heat treatment. The gels formed are opaque, which indicates their particulate structure and suggests a coagulation mechanism. Due to the very low proportion of apo-LDLs (3% of proteins) and the heat-resistance of phosvitin, it can be assumed that HDLs are responsible for gelation of solubilized granules.

Regarding purified LDLs solutions (4% w/v), denaturation starts at 70 °C and gelation from 75 °C. The same solutions heated at 80 °C for 5 minutes form more stable gels than those obtained with bovine serum albumin. In contrast to BSA, LDLs exhibit gelation to heating over a wide pH range. Between pH 6 and pH 9, LDLs form opaque gels, whereas for extreme pH (4–6 and 8–9), the gels are translucent.

LDL heat gelation is controlled by protein denaturation. The first step is LDL destructuration which is favored by protein denaturation. The second step consists of implementation of protein interactions, especially hydrophobic interactions. As regards the gelation of LDLs to freezing, it appears that the dehydration caused by the formation of ice crystals favors the denaturation of the proteins. Then, as for heat gelation, protein-protein interactions allow the establishment of the gel.

1.4. From eggs to egg products

Egg products are processed products obtained from eggs, after at least shell breaking and elimination and heat treatment. They are all intended for human consumption. Other foodstuffs or authorized additives may be partially added to them. They may be either liquid, concentrated, dried, or frozen. This definition leads us to consider two broad categories of egg products: the so-called first processed egg products, corresponding to whole egg, yolk or egg albumen, are proposed in various forms and used as techno-functional ingredients; the so-called specialty egg products result from the production of conventional recipes at an industrial scale and are therefore cooked or pre-cooked products, such as hard-boiled eggs, poached eggs, fried eggs, omelets, beaten eggs, etc.

The egg products industry has experienced significant growth for several years (+ 6.7%/year between 1990 and 2000 in France). Like any agri-food industry, this sector is subjected to a number of regulatory constraints, one of the major objectives of which is the control of hygiene. To this end, requirements are determined for the organization of industrial sites (buildings and flows) and the quality of the raw material: only eggs with a fully developed and flawless shell are usable for the preparation of egg products intended for human consumption; incubated eggs are banned. The content of an egg laid by a healthy hen is, in a very large majority of cases, sterile; and if the shell is entire, it will remain so for a long time, especially when stored at refrigeration temperature.

On the contrary, as soon as the natural protection (egg shell) is eliminated, egg content is inevitably contaminated, since it is impossible to apply a sterile breaking process at an industrial scale. Thus, egg product preservation poses many difficulties. This difficulty is lower regarding egg albumen which is an unfavorable medium for the development of microorganisms, because of the many antibacterial proteins it contains, as well as its alkaline pH and high viscosity. On the contrary, preservation is a huge concern regarding egg yolk and whole egg which are ideal media for microorganism growth.

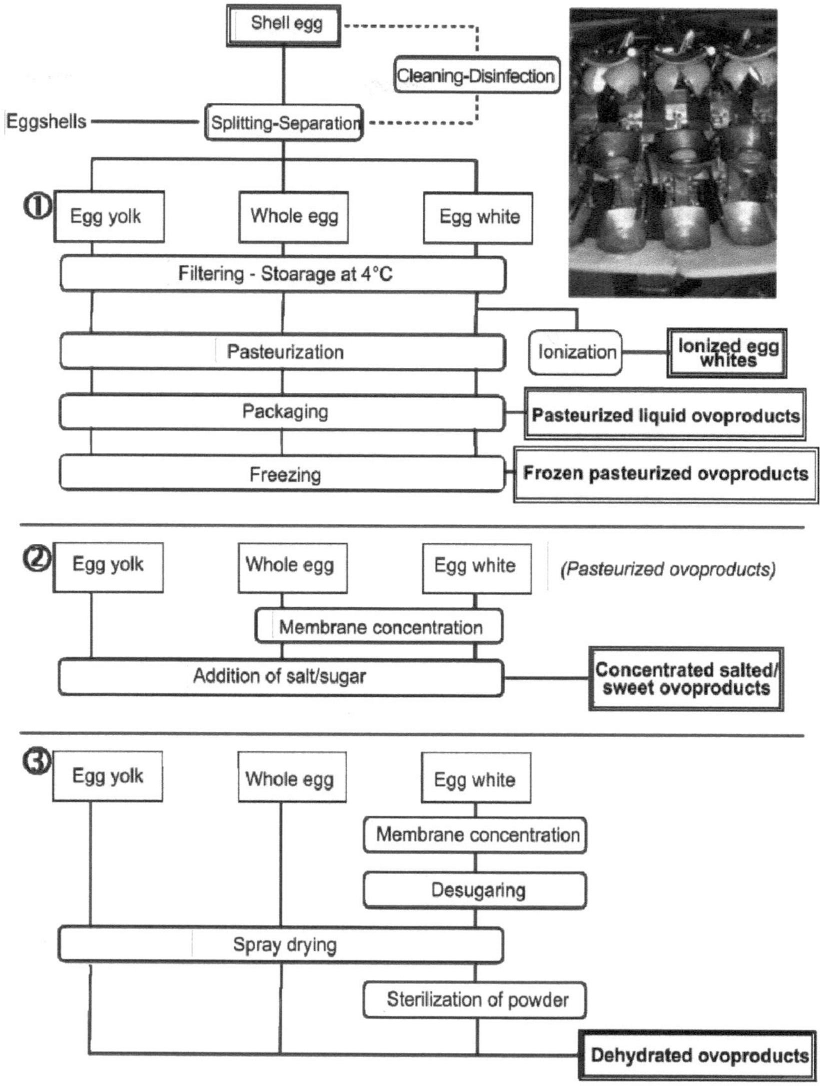

Figure 1.8. *Production diagram of primary processing egg products*

In order to control the microbiological quality of egg products, it is, therefore, necessary to limit their initial contamination, to eliminate all or part of this contamination flora, and to limit or even prevent its development by various means of stabilization. The difficulty that companies have to face is the very high sensitivity of egg proteins towards technological processes used for stabilization, and especially heat-treatments. These treatments result in decreased techno-functional properties, which are particularly damaging in the case of the first processed egg products. The egg product industry now offers a range of products, resulting from various technological schemes, in order to respond as closely as possible to the techno-functional requirements of their customers (Figure 1.8).

1.4.1. *Egg shell decontamination*

Apart from respecting good hygienic practices (cleaning and disinfection of equipment, hygiene of employees, no cross traffic of products), one possible way to limit the egg product contamination is cleaning and disinfection of the egg shells before breaking. Indeed, the elimination of surface flora by brushing, washing, disinfecting, rinsing and drying the eggs can theoretically reduce the microbial load of the egg products, as long as all these operations are carried out very carefully; the quality of the drying step is particularly decisive with regard to the final result. But this is time consuming, costly, and some studies have not found any impact of egg shell washing on the microbiological quality of pasteurized egg products; moreover, the egg washing machines get quickly fouled up. In practice, very few egg product manufacturers wash the eggs before breaking in France, and more generally in Europe.

1.4.2. *Egg breaking and albumen/yolk separation*

The eggs must be broken individually, either mechanically or manually. Specific equipment (so-called "breaking machines") has therefore been developed. The current machines operate at production rates of up to 160,000 eggs per hour and allow an immediate or very slightly deferred separation of albumen and yolk. The products are then filtered to remove the shell fragments (less than 100 mg shell.kg^{-1} of product is required), and then cooled to 4 °C before being sent to storage tanks. Adjustments of dry matter content, addition of salt, sugar or certain authorized additives (guar, xanthan in egg white for example) may be carried out in the storage tanks.

At the end of the breaking and separation steps, the raw products obtained are egg albumen (10 to 11% dry matter, pH between 8.5 and 9.5), egg yolk (42 to 48% dry matter, pH around 6.5), and whole egg (20 to 24% dry matter, pH between 7 and 7.5). Dry matter variations for yolk and whole egg are mainly related to the quality of egg white-yolk separation, depending on the type of machine and the rate of use.

The shells eliminated at this stage constitute a by-product with a relatively high moisture content (nearly 30%), due to egg white adhering onto the shells. This residual moisture poses a risk of important microbial development in this by-product which must therefore be stabilized. The shells are then crushed and centrifuged to separate a part of the adhering egg white and thus to reduce the moisture content by about half. They are then dried in hot-air tunnels before being used, mainly as a limestone amendment for agriculture. A small part of these shells is also intended for animal nutrition as a calcium source, or even for human consumption. In both cases, heat treatments are previously applied in order to guarantee the microbial quality of egg shell powders.

1.4.3. *Egg product stabilization*

1.4.3.1. *Micro-organism destruction: pasteurization*

Among the characteristics of egg proteins, their thermo-sensitivity is undoubtedly the most constraining for manufacturers. The most heat-sensitive egg white proteins are denatured at around 55 °C, and yolk gelation occurs around 70 °C, making egg product sterilization impossible, and limiting pasteurization to time-temperature treatments much lower than those applied on milk for example. However, pasteurization remains the most effective and the most widespread process for microbial decontamination of liquid egg products.

The intensity of pasteurization treatments is determined depending on several factors:

– the nature of the egg product: because of a higher heat-sensitivity, liquid egg white pasteurization is much less intense than whole egg and yolk treatment. Time-temperature combinations can be increased when salt or sugar is previously added to the liquid egg products, without negative consequences on functional properties;

– the desired shelf life: from a few days at 4 °C for liquid egg products in large containers or tankers and intended for food industry, to several weeks (up to 60 days) for liquid egg products in small packs (1 to 2 kg) intended for restaurants or artisans. Longer shelf life (several months at room temperature) can even be obtained for semi-liquid egg products, previously concentrated and with added high concentrations of sugar or salt;

– the desired functionality (especially foaming properties of egg white), the negative effect of heat treatments on the functional properties being proportional to the treatment intensity.

In all cases, the goal is the complete destruction of pathogenic bacteria, in particular *Salmonella*, and the preservation of the product until the end of the shelf life.

The materials used for the liquid egg product pasteurization are conventional plate heat exchangers or tubular heat exchangers, combined with sufficient holding tubes to maintain the desired temperature for the desired time. The standard pasteurization treatment for liquid whole egg and yolk is 66 °C for two to six minutes, whereas it is only 57 °C for 2 to 5 minutes for egg white. Different time–temperature combinations can be applied when using some specific technologies, such as Actijoule® technology (current-conducting tube) which allow higher temperature short time (HTST) treatments (up to 74 °C for 10 to 20 s for whole egg, for example). This type of treatment, followed by an ultra-clean packaging, makes it possible to obtain long lasting egg products.

Faced with the impossibility to apply high heat treatments on liquid egg albumen, the egg product industry has developed a specific process for many years which consists of heating previously dehydrated egg white. The so-called dry-heating treatment has thus become a systematic decontamination process of egg white powder. Dry-heating at around 65 °C for about 10 days ensures the destruction of pathogenic bacteria and a large part of the total flora. Moreover, more severe dry-heating treatments (from 75 to 80 °C for 10 days) also improve the functional properties of egg albumen (Figure 1.9), due to structural protein modifications generated by such treatments. The increase in protein flexibility, partial unfolding, and deamidation, the exposure of reactive groups, in particular the increase in surface hydrophobicity, as well as the lower extent of protein aggregation would explain the higher foaming and gelling properties of strongly dry-heated egg white [KAT 89, KAT 90a, KAT 90b, MAT 01].

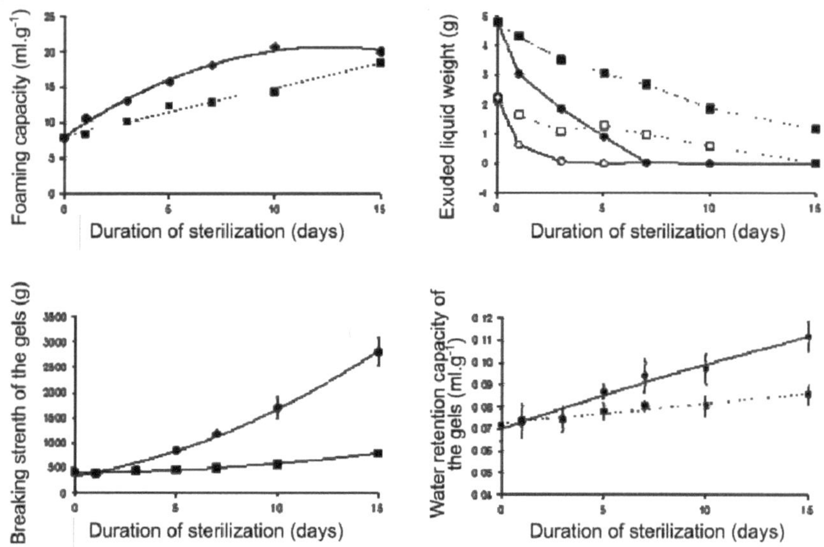

Figure 1.9. *Evolution of foaming and gelling properties of egg white along dry-heating at 67 °C and 75 °C [BAR 03]*

1.4.3.2. *Chemical stability: oxidation*

The consequences of heat treatment of egg products, either in liquid or in powder form, can be the development of oxidative reactions leading to a decrease of product quality, especially nutritional and sensory properties. Due to the high cholesterol content of egg (~400 mg/100 g edible portion) its oxidation in egg and egg products has been extensively studied. Cholesterol oxidation is limited in shell eggs and the oxidation appears to be promoted by further processing [BAR 03]. Concentrations of total cholesterol oxidation products (COPs) in egg products ranged from trace amount up to 200 ppm [HUR 07]. The most predominant COPs identified were 7-ketocholesterol, as β-epoxycholesterol and α-epoxycholesterol [HUR 07]. The amount of COPS in foods can frequently reach 1% of the total cholesterol but occasionally 10%. The degree of oxide formation is related to processing temperature, duration of heating, as well as storage conditions, including packaging [HUR 07]. Moreover the total level of COPs was reported to be higher in spray-dried products than in freeze-dried ones, and to be related to the water content in powders especially during the storage [OBA 06]. As expected, cholesterol oxidation is related to other oxidative reactions occurring in the product and is generally correlated with the lipid unsaturation and with the antioxidant content of the product. Thus the COPs' content in egg yolk powder produced from hens fed with different added fat was fish>flax>sunflower>palm. Supplementation of the diet with tocopherols resulted in a decrease of COPs during storage at room temperature up to four months and after heating [LI 96].

Currently, there is a growing interest in the consumption of foods that have a higher content of n-3 polyunsaturated fatty acids (PUFA) than in conventional foods. As previously mentioned, the enrichment in n-3 PUFA of egg lipids can be accomplished by the supplementation of the hens' diets with fish oil and/or linseed [FRA 12, BAU 00]. However, this enrichment with polyunsaturated fatty acids potentially increases the oxidability of the products, especially during drying or storage. Up to now, in France, the production of dried egg products was limited to the transformation of eggs arising from conventional egg production (standardized diet). Nevertheless, one way to diversify the supply and possibly to enlarge the market of egg products may be to produce n-3-enriched egg powders that can be incorporated further during the processing as food. Moreover, their uses as ingredients in a wide range of food products can contribute to increasing the consumption of n-3 lipids. Nevertheless, the oxidative stability of enriched eggs throughout the egg processing need to be evaluated as well as the impact on the products' quality before being used. Therefore, eggs produced according to the specifications of the "Blue, White, Heart" chain, using extruded linseed as a source of ALA, were prepared as egg products on a pilot plant and their nutritional and sensory properties assessed.

PUFA enrichment and spray-drying temperature do not impair the odour of egg yolk powder. Significant differences between odour of the whole egg powder and egg-yolk powder were highlighted as well as differences between products at the industrial and pilot plant. The odour intensity increased with the time of storage, while unpleasant odour appeared in egg powder being stored at 30 °C [RAN 13].

The fatty acid content, particularly that of polyunsaturated fatty acids, was not significantly modified after the eight months of storage of the powders. This result strengthened the nutritional value of the egg yolk powders. Enriched eggs also contain significantly higher amounts of lutein and zeaxanthin than standard eggs. Although the contents of lutein and zeaxanthin decrease after pasteurization and drying, the amounts finally present in the powders after eight months of storage remain higher in the case of fortified products.

The other remarkable result was the very good oxidative stability of the products. No cholesterol oxidation products have been detected in powders (control or enriched). Lipid oxidation markers (both primary and secondary products), even though they increased as a result of drying and storage, remain at low levels, confirming the results obtained in the studies of sensory properties [MEY 14].

Some hypotheses and further possible research could be outlined. The particular structure of the egg yolk lipids in the lipoproteins could promote their stability, as well as the esterification of the longer PUFA within the phospholipids located at the interface of the lipoproteins. Additionally, the antioxidant activity of egg yolk powder could be investigated as it is a widely used ingredient in the food industry.

1.4.3.3. *Inhibition of micro-organism growth: water activity decrease*

Different stabilization techniques, conventionally used in many agri-food sectors, are applied to egg products to prevent the development of micro-organisms. They all consist in lowering water activity (Aw) below the thresholds, allowing the growth of micro-organisms.

Freezing thus extends the shelf life of egg products up to 24 months. However, it has deleterious consequences for yolk and whole egg, namely gelation which presents a considerable increase of viscosity of the products after thawing. The effects are lower on egg white. This method of stabilization however remains little used in practice since thawing durations are quite long with high risks of microbial development during this step. In all cases, it can only concern small and medium packagings, that is to say below 20 kg.

Concentration combined with the addition of sugar and/or salt at high concentrations is used to obtain egg products which are stable for several months at room temperature. Egg white is usually concentrated by ultrafiltration or reverse osmosis, up to 30% dry matter (threefold concentration); sugar or salt are then added after concentration, and before pasteurization. A significant decrease in Aw can thus be achieved (Aw 0.88); yet despite this, it does not enable long-term storage at room temperature. Ultrafiltration can also be used for whole egg concentration, but vacuum evaporation is more common; it is then performed after sugar or salt addition and pasteurization. For example, twofold concentrated whole egg (about 48% dry matter) added with 50% sugar or 9% salt has an Aw of 0.80 and 0.85, respectively; such Aw levels enable storage at room temperature. On the contrary, egg yolk is never concentrated, because it is naturally highly concentrated (higher than 44% dry matter); the addition of 12% salt or 50% sugar is then enough to achieve Aw of 0.85.

Drying makes it possible to produce powdered egg products and is the simplest, longest, safest and most flexible form of preservation. These numerous advantages explain the development of egg powders all over the world, especially regarding egg white and whole egg which are widely exchanged at an international level. The technology used is spray-drying performed in horizontal (box-dryers) or vertical dryers. Whole egg and yolk are spray-dried after pasteurization. On the contrary, regarding egg white, pasteurization is performed at the very end of the process, on the powder: this is the dry-heating mentioned in section 1.4.3.1. Moreover, prior to drying egg white, a desugarization step is necessary in order to avoid the development of the Maillard reaction (non-enzymatic browning) during the drying itself, but especially during dry-heating. This desugarization consists in removing glucose naturally present in egg white (about 0.5 $g.l^{-1}$). It is carried out either by controlled fermentation (yeasts or bacteria) or by enzymatic means using a combination of glucose oxidase and catalase enzymes.

1.4.4. *Specialty egg products*

Specialty egg products, most of which are intended for food service, artisans (pastry chefs, caterers, etc.) or specific food industries (sandwiches, salads, etc.), represent only 10% of total egg products in France, including about 80% for hard-boiled eggs only.

The industrial process for hard-boiled eggs prevails over the steps implemented at household level. The cooking step is carried out either by immersion in hot water (98–100 °C) or in steam. After cooling by immersion in cold water, egg shells are fragmented and then eliminated (peeling step). This latter step is undoubtedly one of the most difficult to control on an industrial scale, explaining the very large number of processes and materials that have been developed and are currently proposed by the equipment suppliers.

Fragmentation of the shells is generally achieved by shocks caused between cooked eggs. For egg peeling, systems based on rollers, bars, or "fingers" of rubber are used, a jet of water under high pressure facilitates this operation. However, a visual check and a manual step of removing residual shell fragments are essential at the end of the line. The hard-boiled eggs are then rinsed, drained and packaged, either in brine or in plastic film, possibly in a modified atmosphere. The two key points that manufacturers must control in the process are:

– the level of cooking: the yolk must be completely cooked, while avoiding any overcooking which could lead to discoloration of the yolk surface due to the formation of iron sulfide. Hydrogen sulfide from the egg white reacts with the iron from the yolk, according to a chemical reaction whose kinetics is totally dependent on the temperature;

– the ease of peeling: the facility for separation of shell fragments from the cooked egg white is clearly correlated with the albumen pH before cooking; it must be higher than 8.5. This criterion therefore excludes the use of very fresh eggs for hard-boiled egg production.

Other processes have been developed to produce, on an industrial scale, specialty egg products such as poached, fried, scrambled eggs, omelets, etc. Despite the apparent simplicity of these products, many technical difficulties are hidden behind the processes that make them possible.

Furthermore, it is possible to extract a large number of egg constituents, in particular proteins, on a laboratory or pilot scale. These products are available for many suppliers of biochemicals. But only lysozyme, which is used both in food and pharmaceutical industries for its antibacterial properties, represents significant tonnages worldwide. Egg white lysozyme is extracted by cation exchange chromatography.

1.5. Specific industrial applications

The gradual industrialization of traditional recipes led to formulation adjustments in order to optimize processes, reduce costs and increase products' shelf life. In many sectors, egg products have been specifically developed to answer customer requirements; in other cases, innovative egg products led to a modification of existing practices for food production. Thus, egg product companies are focused on the market of sauces on the one hand, and on sweet dishes on the other regarding egg yolk, with original and sophisticated technical solutions. The recipes of liquid egg albumen have multiplied in the world of sweet products, mainly for applications where foam albumen is used. Conversely, for salty products, egg albumen powders are increasingly used, instead of liquid egg white, because of much better binding and gelling properties. Whole egg, except for a few applications, remains a generic product for which no specific optimization has been developed.

1.5.1. *Egg yolk in sauces*

Egg yolk added with 8–12% salt, and possibly 2 to 10% sugar, has become a must for the production of cold emulsified sauces such as mayonnaise. European guidelines for good practice impose on mayonnaise recipes 5 to 6% egg yolk, for 65 to 77% oil. The formulations thus remain diversified, for cultural reasons, and above all because the uses vary.

Salted egg yolk remains the reference emulsifier for sauce applications. The increase in ionic strength changes the granule structure and promotes emulsification at acidic pH. This change in structure results in an increase of the viscosity of egg yolk, from 20 mPa.s at 4 °C for a natural yolk, to about 200 mPa.s for a 10% salted yolk. The high viscosity of salted yolk influences the rheological characteristics of the aqueous phase of the mayonnaise, favoring its stability by limiting coalescence. It is noticeable that the viscosity of egg yolk naturally exhibits a high variability which has not been extensively investigated; the relative weight of various factors such as the genetics of the hens, the fatty acid composition or the influence of shell egg storage is still not well known.

The texture of mayonnaise is an essential criterion for the hedonic dimension of this product. The rheological parameters primarily depend on the size of the oil droplets. The recent evolution of shear devices has made it possible to obtain, for a given formulation, increasingly fine droplets with a smaller dispersion of size. But it is also possible to decrease this size by increasing the proportion of yolk. The salt or sugar content has a lower impact on 65% oil emulsions. Particularly severe pasteurization of natural yolk (68 °C – 11 min) also results in a reduction of the size of mayonnaise droplets by 40% compared to unpasteurized yolk [GUI 06]. Thus, the technological parameters applied on egg yolk have an impact on the texture of the final mayonnaise. The tendency to reduce the oil content of sauces, observed for several years, will certainly induce technological adjustments on egg yolk, to limit the use of complementary ingredients.

The presence of salt and sugar protects the color and aroma of egg yolk during heat treatments. Moreover, salted egg yolks effectively tolerate a wide range of heat treatments between 65 and 77 °C [CAM 05], given that a standard treatment of 64.4 °C for 2.5 minutes is sufficient to obtain 14 decimal reductions of *Salmonella* in 10% salted yolk. However, pasteurization technologies must be adapted to take account on the one hand, of the high viscosity of these products, and on the other hand, the increased heat-resistance of micro-organisms when Aw decreases. Such formulas can be stored for several months at room temperature, even if storage at –5 °C would be preferable regarding volatile compound content and emulsifying properties for a long storage period (six months) [YAN 09].

An original treatment of yolk, developed in the 1970s, makes it possible to obtain heat-stable mayonnaise. It involves hydrolyzing yolk phospholipids by using a phospholipase enzyme [DUT 81]. This results in an increase of the emulsifying properties of yolk. Such modified yolk has contributed to improving the microbiological safety of mayonnaise which could be packaged at high temperature, even pasteurized.

1.5.2. *Egg yolk in sweet products*

Egg yolk is also a traditional ingredient for the preparation of desserts. It is used at 2 to 10% to bring color and smoothness, or to limit sugar crystallization in recipes such as custards, pastry cream, ice creams or egg creams. It is sometimes used at higher percentages, in specialties such as “lemon curd”, Portuguese candy exclusively made from egg yolk and sugar, or in alcoholic beverages such as the Dutch “advocaat”.

Diluted egg yolk has better foaming properties than whole egg, which is rarely used except for “sabayon”. In pastry, it is also usual to “blanch” egg yolk, by incorporating air into the yolk before adding sugar and mixing it with the dough. This makes it possible to increase the rising of the dough during cooking, through the evaporation of water which develops the pre-existing air bubbles.

For all of these applications, egg yolk is offered either natural or sweet at 10%. Sugar addition is often an interesting solution since it protects the original organoleptic qualities of yolk against heat treatments, during pasteurization or when incorporated into hot milk, for example. Up to 20 to 50% sugar is sometimes added to yolk to accentuate this protective effect, as long as sugar only slightly increases yolk viscosity which remains equivalent to that of natural yolk. Yolk dilution with whole egg is required for the products to be gelled after cooking. Tailor-made formulations are easy to implement regarding sweet products, since they do not require technologies other than those necessary for the preparation of yolk including 10% sugar. However, pasteurization time and temperature should be adjusted according to the sugar content, since micro-organisms' resistance increases when Aw decreases. Thus, in order to obtain 5 decimal reductions of *Salmonella*, 61.1 °C for 3.5 minutes is efficient for natural yolk, whereas 63.3 °C for 3.5 minutes is necessary for yolk with 10% sugar. In return, the Aw of egg yolk containing 50% sugar and 73% dry matter is around 0.85; this product thus offers an excellent microbiological stability after pasteurization. Although rarer, 10% sugar egg yolk is sometimes frozen. Indeed, sugar and glycerol are excellent cryo-protective agents to limit the denaturation observed in the case of natural yolk after freezing [TEL 98].

Egg yolk intended for the preparation of fresh desserts requires extremely low residual floras after pasteurization. Precautions must therefore be taken throughout the manufacturing chain to limit the presence of spore-forming thermophilic flora that could withstand heat treatment. Indeed, some of these microorganisms are psychrotrophic and can therefore alter the quality of fresh finished products during their storage. Neither the increase in heat treatments, nor the decrease in Aw is a sufficient response to these thermoresistant and ubiquitous floras. Only specific technical solutions aimed at limiting the introduction and the growth of biofilms have made it possible to produce these microbiologically demanding egg products. These products have contributed to the expansion and diversification of fresh egg-based desserts in Western Europe.

1.5.3. *Egg albumen for foaming products*

Egg white is incorporated into a multitude of recipes for its foaming properties. Egg white foams, natural or including 10 to 30% sugar, are thus incorporated without cooking in sweet sauces (chocolate mousse) or in cake dough (Savoy biscuit, sponge finger) to lighten them. In cakes, egg white also contributes to the firmness of the product after cooking, because of its gelling properties. Egg white foams with a low sugar content are used to prepare "floating islands" after cooking. Other recipes use egg white foams containing 40 to 80% sugar, such as macaroons, Italian or French meringues, or nougats. In this case, natural egg white is whipped to obtain a foam in which powdered sugar or hot sugar syrup is gradually incorporated. For example, Italian meringues are prepared with a hot sugar syrup at 121 °C which cooks the egg white at the end of whipping; these meringues are used to lighten pastry fillings such as butter cream or mousseline cream, or to make decorations on lemon pies for example. For all other applications, egg white foam is stabilized by oven cooking.

In addition, egg white foam is used increasingly to lighten salty preparations, especially festive products (vegetable and fish mousses). Again, there are many methods to incorporate egg white. In all cases, it remains difficult to model the culinary qualities of egg white in the complex formulas in which it is incorporated, from simple functional tests conventionally performed on pure egg white.

By comparison with a manually separated egg white, mechanically separated egg white (industrial egg white) results in lower foam volume and stability: breaking step, numerous air-liquid interfaces created, shear and pasteurization contribute to the decrease of foaming properties [LEC 05]. Then, although the use of pasteurized egg white to prepare uncooked products such as chocolate mousse is a significant improvement for food safety, compared to the traditional use of raw egg white, it is, however, complex to practically maintain the excellent foaming properties of egg white when previously pasteurized.

The lower stability of egg white foam resulting from pasteurization can be made up for through the addition of hydrocolloids (guar gum, xanthan, carrageenans, pectins, etc.) before heat treatment. However, these additives generally reduce foam capacity [IBA 07]. The firmness of the foams thus obtained ultimately limits the volume losses when mixed with a sweet sauce, making it possible to produce quality chocolate mousses, for example. Moreover, in the case of floating islands, higher foam expansion is obtained during cooking in the presence of hydrocolloids, and the volume yields of the finished products are also satisfactory.

Sugar addition in egg white offers additional possibilities to increase the efficiency of pasteurization while maintaining foaming properties. Thus, the foaming capacity of 12% sugar egg white increases when pasteurization increases from 60 °C-2 minutes up to 64 °C-2 minutes; however, the foam stability slightly decreases [RAI 07].

When high foaming properties are required, "high foam" egg white powders can be also used. These egg white powders are obtained by applying dry-heating treatments specifically designed to control these properties (4.3.1).

1.5.4. *Whole egg uses*

Almost 70% of pasteurized egg products are used as whole egg in traditional sectors such as pastry, baking and catering. These egg products replace the shell eggs in traditional recipes, with a functional expectation similar to that of raw egg. For most applications, the formulated products are cooked, justifying that the microbiological requirements on the egg product only concern *Salmonella* and mesophilic flora. Fresh food products such as salted pies, quenelles or fresh pasta are more sensitive to heat-resistant psychrotrophic bacteria. That is why some manufacturers have requirements with regard to these bacteria, in order to better control the shelf life of their own products.

The whole egg is a versatile ingredient in pastry and biscuits. Its role in cakes, before cooking, mainly depends on the process of air incorporation. In whipped dough (sponge cake), whole egg is used for its foaming properties in the presence of sugar, before addition of the other ingredients. For risen dough (brioche, cake), the emulsifying and film-forming properties of the whole egg contribute to retaining the gases produced by yeasts or the baking powder to form a characteristic alveolar structure. Whole egg also acts as a binder in shortbread biscuits, for example. When cooked, whole egg interacts with other ingredients such as flour to create a gelled network. The rate of gelation is decisive for the height of whipped or risen dough. Finally, either natural or formulated, and applied as a thin layer on the surface of pastries, whole egg results in a bright and golden appearance after cooking.

The pasteurization conditions of whole egg have an influence on the finished products in the case of risen dough. However, very few studies have investigated these issues, and would deserve to be extended to all cakes in which egg has a determining role. Moreover, in many cases, the addition of emulsifiers, texturizing agents, and flour treatment agents equalize the initial foaming properties of the egg products. The method of incorporating eggs would also affect the batter density before cooking in the manufacture of sponge fingers, depending on whether whole egg is premixed with water and sugar before flour addition, or mixed with the whole recipe [ALL 06].

Whole egg is used in dry pasta, especially to provide a pleasant color. But it is now limited to regional specialties, for which a precise color of the egg product is often required. Whole egg also remains a major ingredient in fresh pasta. Indeed, this application requires pasteurization in the packaging in order to obtain a shelf life compatible with distribution networks. Thus, whole egg gives an excellent thermostability to fresh pasta. Moreover, the egg white/yolk ratio can be modified in order to increase the firmness (when this ratio increases), or to favor swelling during cooking (when it decreases) [ALA 09]. The relationship between the dough quality and the intensity of egg pasteurization could also be established.

The qualitative criteria described above for fresh pasta production, namely the yolk color, the egg white to yolk ratio and the pasteurization intensity, are also the main factors that should be considered for the optimization of salty product preparation (salty pies, quenelles, etc.).

1.6. Bibliography

[ALA 09] ALAMPRESE C., CASIRAGHI E., ROSSI M., "Modeling of fresh egg pasta characteristics for egg content and albumen to yolk ratio", *J. Food Eng.*, vol. 93, no. 3, pp. 302–307, 2009.

[ALL 06] ALLAIS I., EDOURA-GAENA R-B., GROS J-B. *et al.*, "Influence of egg type, pressure and mode of incorporation on density and bubble distribution of a lady finger batter", *J. Food Eng.*, vol. 74, no. 2, pp. 198–210, 2006.

[ALU 98] ALUKO R.E., KEERATIURAI M., MINE Y., "Competitive adsorption between egg yolk lipoproteins and whey proteins on oil-in-water interfaces", *Colloids Surf. B Biointerfaces*, vol. 10, no. 6, pp. 385–393, 1998.

[ANT 13] ANTON M., "Egg yolk: structures, functionalities and processes: Egg yolk: structures, functionalities and processes", *J. Sci. Food Agric.*, vol. 93, no. 12, pp. 2871–2880, September 2013.

[ANT 00] ANTON M., DENMAT M.L., GANDEMER G., "Thermostability of hen egg yolk granules: Contribution of native structure of granules", *J. Food Sci.*, vol. 65, no. 4, pp. 581–584, 2000.

[ANT 97] ANTON M., GANDEMER G., "Composition, solubility and emulsifying properties of granules and plasma of egg yolk", *J. Food Sci.*, vol. 62, no. 3, pp. 484–487, 1997.

[BAR 03] BARON F., NAU F., GUERIN-DUBIARD C. *et al.*, "Effect of dry heating on the microbiological quality, functional properties, and natural bacteriostatic ability of egg white after reconstitution", *J. Food Prot.*, vol. 66, no. 5, pp. 825–832, 2003.

[BAU 00] BAUCELLS M., CRESPO N., BARROETA A. *et al.*, "Incorporation of different polyunsaturated fatty acids into eggs", *Poult. Sci.*, vol. 79, no. 1, pp. 51–59, 2000.

[CAM 05] CAMPBELL L., RAIKOS V., EUSTON S.R., "Heat stability and emulsifying ability of whole egg and egg yolk as related to heat treatment", *Food Hydrocoll.*, vol. 19, no. 3, pp. 533–539, 2005.

[CRO 02] CROGUENNEC T., NAU F., BRULE G., "Influence of pH and salts on egg white gelation", *J. Food Sci.*, vol. 67, no. 2, pp. 608–614, 2002.

[DAM 98] DAMODARAN S., ANAND K., RAZUMOVSKY L., "Competitive adsorption of egg white proteins at the air-water interface: Direct evidence for electrostatic complex formation between lysozyme and other egg proteins at the interface", *J. Agric. Food Chem.*, vol. 46, no. 3, pp. 872–876, 1998.

[DEN 99] DENMAT M., ANTON M., GANDEMER G., "Protein denaturation and emulsifying properties of plasma and granules of egg yolk as related to heat treatment", *J. Food Sci.*, vol. 64, no. 2, pp. 194–197, 1999.

[DOI 93] DOI E., "Gels and gelling of globular proteins", *Trends Food Sci. Technol.*, vol. 4, no. 1, pp. 1–5, 1993.

[DUT 81] DUTILH C.E., GROGER W., "Improvement of product attributes of mayonnaise by enzymic hydrolysis of egg yolk with phospholipase A2", *J. Sci. Food Agric.*, vol. 32, no. 5, pp. 451–458, 1981.

[DYE 93] DYER-HURDON J., NNANNA I., "Cholesterol content and functionality of plasma and granules fractionated from egg yolk", *J. Food Sci.*, vol. 58, no. 6, pp. 1277–1281, 1993.

[FRA 12] FRAEYE I., BRUNEEL C., LEMAHIEU C. *et al.*, "Dietary enrichment of eggs with omega-3 fatty acids: A review", *Food Res. Int.*, vol. 48, no. 2, pp. 961–969, 2012.

[GUE 10] GUERIN-DUBIARD C., ANTON M., GAUTRON J. *et al., Composition de l'œuf, Science et technologie de l'œuf. Volume 2 : de l'œuf aux ovoproduits*, Lavoisier, Paris, 2010.

[GUI 06] GUILMINEAU F., KULOZIK U., "Impact of a thermal treatment on the emulsifying properties of egg yolk. Part 1: Effect of the heating time", *Food Hydrocoll.*, vol. 20, no. 8, pp. 1105–1113, 2006.

[HUR 07] HUR S.J., PARK G.B., JOO S.T., "Formation of cholesterol oxidation products (COPs) in animal products", *Food Control*, vol. 18, no. 8, pp. 939–947, August 2007.

[IBA 07] IBANOGLU E., ERÇELEBI E.A., "Thermal denaturation and functional properties of egg proteins in the presence of hydrocolloid gums", *Food Chem.*, vol. 101, no. 2, pp. 626–633, 2007.

[KAS 10] KASSIS N., DRAKE S.R., BEAMER S.K. *et al.*, "Development of nutraceutical egg products with omega-3-rich oils", *LWT - Food Sci. Technol.*, vol. 43, no. 5, pp. 777–783, June 2010.

[KAT 90a] KATO A., IBRAHIM H.R., WATANABE H. *et al.*, "Structural and gelling properties of dry-heated egg white proteins", *J. Agric. Food Chem.*, vol. 38, no. 1, pp. 32–37, 1990.

[KAT 90b] KATO A., IBRAHIM H.R., TAKAGI T. *et al.*, "Excellent gelation of egg white preheated in the dry state is due to the decreasing degree of aggregation", *J. Agric. Food Chem.*, vol. 38, no. 9, pp. 1868–1872, 1990.

[KAT 89] KATO A., IBRAHIM H.R., WATANABE H. *et al.*, "New approach to improve the gelling and surface functional properties of dried egg white by heating in dry state", *J. Agric. Food Chem.*, vol. 37, no. 2, pp. 433–437, 1989.

[KIO 89] KIOSSEOGLOU V., *Egg Yolk*, Elsevier, London, 1989.

[KIT 88] KITABATAKE N., SHIMIZU A. *et al.*, "Preparation of Transparent Egg White Gel with Salt by Two-step Heating Method", *J. Food Sci.*, vol. 53, no. 3, pp. 735–738, 1988.

[LEC 05] LECHEVALIER V., CROGUENNEC T., PEZENNEC S. *et al.*, "Evidence for synergy in the denaturation at the air–water interface of ovalbumin, ovotransferrin and lysozyme in ternary mixture", *Food Chem.*, vol. 92, no. 1, pp. 79–87, 2005.

[LEC 03] LECHEVALIER V., CROGUENNEC T., PEZENNEC S. *et al.*, "Ovalbumin, Ovotransferrin, Lysozyme: Three Model Proteins for Structural Modifications at the Air-Water Interface", *J. Agric. Food Chem.*, vol. 51, no. 21, pp. 6354–6361, 2003.

[LI 96] LI S., CHERIAN G., SIM J., "Cholesterol oxidation in egg yolk powder during storage and heating as affected by dietary oils and tocopherol", *J. Food Sci.*, vol. 61, no. 4, pp. 721–725, 1996.

[LI 89] LI-CHAN E., "Biochemical basis for the properties of egg white", *CRC Crit. Rev. Poult. Biol.*, vol. 2, pp. 21–58, 1989.

[MAR 03] MARTINET V., SAULNIER P., BEAUMAL V. *et al*, "Surface properties of hen egg yolk low-density lipoproteins spread at the air–water interface", *Colloids Surf. B Biointerfaces*, vol. 31, no. 1, pp. 185–194, 2003.

[MAR 02] MARTINET V., BEAUMAL V., DALGALARRONDO M. *et al.*, "Emulsifying properties and adsorption behaviour of egg yolk lipoproteins (LDL and HDL) in oil-in-water emulsions", *Food Emuls. Dispers.*, pp. 103–116, 2002.

[MAT 01] MATSUDOMI N., TAKAHASHI H., MIYATA T., "Some structural properties of ovalbumin heated at 80 C in the dry state", *Food Res. Int.*, vol. 34, no. 2, pp. 229–235, 2001.

[MEY 14] MEYNIER A., LEBORGNE C., VIAU M. *et al.*, "n-3 fatty acid enriched eggs and production of egg yolk powders: An increased risk of lipid oxidation?", *Food Chem.*, vol. 153, pp. 94–100, 2014.

[MIN 00] MINE Y., KEERATIURAI M., "Selective displacement of caseinate proteins by hens egg yolk lipoproteins at oil-in-water interfaces", *Colloids Surf. B Biointerfaces*, vol. 18, no. 1, pp. 1–11, 2000.

[MIZ 85] MIZUTANI R., NAKAMURA R., "Physical state of the dispersed phases of emulsions prepared with egg yolk low density lipoprotein and bovine serum albumin", *J. Food Sci.*, vol. 50, no. 6, pp. 1621–1623, 1985.

[NEI 16] NEIJAT M., OJEKUDO O., HOUSE J., "Effect of flaxseed oil and microalgae DHA on the production performance, fatty acids and total lipids of egg yolk and plasma in laying hens", *Prostaglandins Leukot. Essent. Fat. Acids PLEFA*, vol. 115, pp. 77–88, 2016.

[OBA 06] OBARA A., OBIEDZIŃSKI M., KOLCZAK T., "The effect of water activity on cholesterol oxidation in spray-and freeze-dried egg powders", *Food Chem.*, vol. 95, no. 2, pp. 173–179, 2006.

[POW 95] POWRIE W.D., NAKAI S., "The chemistry of eggs and egg products", *Egg Sci. Technol.*, vol. 4, pp. 105–75, 1995.

[RAI 07] RAIKOS V., CAMPBELL L., EUSTON S.R., "Effects of sucrose and sodium chloride on foaming properties of egg white proteins", *Food Res. Int.*, vol. 40, no. 3, pp. 347–355, 2007.

[RAN 13] RANNOU C., TEXIER F., MOREAU M. *et al.*, "Odour quality of spray-dried hens' egg powders: The influence of composition, processing and storage conditions", *Food Chem.*, vol. 138, nos 2–3, pp. 905–914, June 2013.

[SHE 79] SHENTON A.J., Membrane composition and performance of food emulsions, PhD thesis, King's College London, 1979.

[SIN 12] SINGH V., PATHAK V., AKHILESH K.V., "Modified or enriched eggs: A smart approach in egg industry: A", *Am. J. Food Technol.*, vol. 7, no. 5, pp. 266–277, 2012.

[SUR 01] SURAI P., SPARKS N., "Designer eggs: from improvement of egg composition to functional food", *Trends Food Sci. Technol.*, vol. 12, no. 1, op. 7–16, 2001.

[TAH 17] TAHERGORABI R., JACZYNSKI J., "Chapter 26 – Nutraceutical Egg Products", in HESTER P.Y. (ed.), *Egg Innovations and Strategies for Improvements*, Academic Press, San Diego, 2017.

[TEL 98] TELIS V., KIECKBUSCH T., "Viscoelasticity of frozen/thawed egg yolk as affected by salts, sucrose and glycerol", *J. Food Sci.*, vol. 63, no. 1, pp. 20–24, 1998.

[THA 94] THAPON J.L., BOURGEOIS W.G., *L'Oeuf et les Ovoproduits*, Lavoisier, Paris, 1994.

[YAN 09] YANAGISAWA T., WATANUKI C., ARIIZUMI M. *et al.*, "Super Chilling Enhances Preservation of the Freshness of Salted Egg Yolk During Long-Term Storage", *J. Food Sci.*, vol. 74, no. 2, 2009.

2

Spoilage of Egg Products

2.1. Introduction

The microorganisms responsible for spoilage do not belong to a single taxonomic group. Bacteria, molds and yeasts can be the cause of alterations. The variety of food microorganisms depends on different factors. The first is relative to the nature of the food, for example, the fact that fish products may be contaminated by water, vegetables by soil and animal products by the intestinal tract of the animal from which the product is derived. The second factor relates to the intrinsic properties of the food, such as its chemical composition, pH, redox potential and/or water availability (estimated by water activity or w_a). The third factor is related to the process of transformation and stabilization that the food product may have undergone. So, as it should be, the balance of the microorganisms can be modified by stabilization treatments (heat treatment, drying, acidification, addition of preservatives, cold storage or modified atmospheres or a combination of one or more of these treatments).

Chapter written by Sophie JAN, Florence BARON, Rémy COAT and Olivier GONÇALVES

Microbial spoilage of food may include the formation of visible colonies, the release of gas, the onset of food pigmentations, and/or discolorations and/or changes in texture and/or odor. The molecules produced are dependent on the substrates available in the food and on the metabolism of the microbial community. Food spoilage is thus a very heterogeneous phenomenon and the nature of the molecules produced can be very varied. They can include volatile or nonvolatile compounds, organic acids, alcohols, ketones, amines, esters, sulfur compounds, pigments, or other molecules that can render the food unfit for consumption.

It must be emphasized that the spoilage of shell eggs causes macroscopic changes (odor and/or color and/or viscosity) which make them unlikely to be exploited by the egg product industry. The focus will therefore be on bacteria that are not very competitive in the egg but which are selected by industrial pasteurization and refrigeration processes. We will focus on the spoiling activities of bacteria that can withstand pasteurization temperatures and cause spoilage during refrigerated storage. A large number of bacteria known to be involved in shell egg spoilage will be described but are unlikely to be involved in the spoilage of egg products due to their heat sensitivity, as long as the pasteurization schedules are optimized and hygiene is excellent after pasteurization. *Salmonella Enteritidis* is the main pathogen involved in foodborne illnesses in developed countries and eggs are the most common food product that causes these infections [EFS 14a]. This type of bacteria is very competitive in the sector of egg production because of its ability to infect eggs during their formation in the genital tract of contaminated hens and of its ability to survive in egg white [BAR 16]. The most vulnerable people are pregnant women, infants and the elderly or immunocompromised individuals. However, the occurrence of this pathogen has decreased considerably in Europe, both in farms and in eggs, thanks to the establishment of monitoring and vaccination programs [EFS 14a] and the pasteurization of egg products which makes it possible to control the risk. This microorganism will not be further mentioned in this chapter.

We will see how an egg product, such as the whole egg, provides all the nutrients necessary for bacterial growth and the expression of adapted hydrolytic activities. This chapter aims to describe the specificities of hydrolytic enzymes potentially expressed by bacteria that come into play in the spoilage of egg products and the main classes of molecules in the egg: proteases/proteins, lipases/triglycerides, phospholipases/phospholipids. The molecules resulting from these bacterial hydrolyses will be listed as well as the end products of the metabolic pathways of these bacteria. Finally, signaling molecules supporting quorum-sensing will be described as potentially interesting targets in the context of research into markers of bacterial spoilage.

2.2. Bacteria involved in shell egg spoilage

According to [BOA 65], the initial signs of bacterial spoilage of the contents of a shell egg are due to the development of chromogenic bacteria or the activity of bacterial hydrolytic enzymes. This type of event is highly dependent on hygiene conditions during egg production, handling, transportation, and of course, egg storage duration and temperatures. The density and nature of the shell flora as well as the storage and handling conditions are indeed key factors in the microbial invasion of the egg. The shell flora is varied. The average level of contamination reported in the literature is around 4.5 $\log_{10}$ bacteria per shell [EFS 14a]. Gram-positive bacteria dominate, such as those belonging to the genera *Micrococcus*, *Staphylococcus*, *Steptococcus*, *Aerococcus* and, to a lesser extent, *Bacillus*. The literature also refers to Gram-negative bacteria, including *Salmonella* but also bacteria of the genera *Escherichia*, *Alcaligenes* and *Pseudomonas*. The bacteria described as responsible for shell egg spoilage are primarily the bacteria present on the surface of the egg that are able to penetrate its natural defense mechanisms, which include the cuticle, the shell, the shell membranes and the vitelline membrane. The development of this flora, if it has managed to cross one or more of these barriers, is dependent on the nutrients available and the storage temperature of the egg. The growth is optimal in egg yolk (rich in nutrients, physiological pH) while it is hindered in the white (lack of essential elements including iron, an alkaline pH and a wide array of molecules with antibacterial activity). Gram-positive bacteria, though most prevalent on the shell, also have the weakest abilities to penetrate the natural defense mechanisms of the egg. Gram-negative bacteria, mainly *Salmonella*, which are mobile and less nutritionally demanding, are the bacteria most likely to penetrate all barriers and multiply [BOA 77, BOA 95, HAR 63, GRI 01]. According to [BOA 94, 00], a period of 7 to 20 days at room temperature is required for the microorganisms present on the shell to be detected in the contents of the egg.

The literature focusing on the bacteria responsible for the spoilage of shell eggs is dated and refers to production conditions that are out of date. Spoiled eggs are described as harboring a mixed flora, consisting mainly of Gram-negative bacteria. Gram-positive bacteria are less common [EFS 14a]. The most representative bacteria belong to the genera *Pseudomonas*, *Proteus* and *Alcaligenes*. Less prevalent genera include *Flavobacterium*, *Cyophaga*, *Aeromonas*, *Serratia*, *Bacillus*, *Arthrobacter*, *Micrococcus* and *Streptococcus* [EFS 14a, BOA 00]. Molds of the genera *Penicillium*, *Alternaria* and *Mucor* have also been identified in spoiled eggs [EFS 14a]. According to [HAI 38], the number of marketed spoiled eggs did not exceed 5% in the 1930s. The development of new breeding and distribution practices has reduced the contamination by aerobic bacteria, including enterobacteria, *Escherichia coli,* yeasts and molds [MUS 05].

2.3. Bacteria involved in whole egg spoilage

The flora of an egg product such as whole egg may come from (1) the contents of the egg if it is contaminated, (2) the laying environment (herd infections, chicken feces, dust, food, etc.) and storage (in packing centers or in industry), due to the unavoidable transfer of some of the microorganisms present on the shells to the egg product during the industrial breaking step or (3) the transformation environment (premises, equipment). With regard to the raw material, a certain number of precautions must be taken when handling and storing the eggs. Indeed, the weakening of the cuticle and the shell should be avoided, especially for eggs whose shell is heavily contaminated. Care must be taken to avoid water condensation on the shell and abrupt changes in temperature during storage. The loss of weight of the egg over time, due to the increase in volume of the inner tube, the separation of the shell membranes and the rupture of the inner shell membrane, also pose a contamination risk of the internal compartments of the egg, upstream of the breakage. Similarly, the exchange of water and minerals between the yolk and the egg white can cause a decrease in the viscosity of the egg white and the weakening of the vitelline membrane over time, especially when temperatures are not controlled. Egg storage conditions must therefore be strictly controlled as they have a definite impact on the microbiological quality of the egg product. The risks of this include (1) the egg allowing microorganisms to enter its internal compartments upstream of the breakage and (2) the transfer of microorganisms from the shell to the egg product at the time of breaking.

[MAC 82] show that Gram-negative bacteria dominate in spoiled raw whole egg after storage at different temperatures (30, 25, 20, 10 and 5°C). The dominant species identified are *Acinetobacter calcoaceticus*, *Aeromonas hydrophyla*, *Bacillus cereus*, *Citrobacter freundii*, *Enterobacter aerogenes*,

Enterobacter cloaca, *Escherichia coli*, *Klebsiella pneumoniae*, *Proteus vulgaris*, *Serratia marcescens*, *Pseudomonas putida*, *Salmonella typhymurium*, *Streptococcus faecalis*, *Septrococcus lactis* and *Vibrio metschnikovii*. Enterobacteriaeae dominate at 20–30°C, whereas Pseudomonadaceae and Vibrionaceae dominate at 20°C and 10°C. At 5°C, Pseudomonadaceae dominate. It is generally accepted that less than 1% of the bacteria present in raw whole egg survive pasteurization. These are Gram-positive coccior bacteria belonging to the genera *Alcaligenes*, *Proteus*, *Escherichia*, *Flavobacterium*, or *Bacillus* [CUN 77]. A study carried out in 2006 in Western France analyzed the evolution of the flora of industrial whole egg products collected after breakage (raw whole egg) or pasteurization (pasteurized whole egg) [PRO 07]. The results clearly show that if mesophilic and potentially pathogenic microorganisms are present in raw, whole egg, they are destroyed by pasteurization. This is particularly the case with *S. Enteritidis*. However, bacteria that are uncompetitive in egg invasion prevail in an egg product such as whole egg. These bacteria are mainly Gram-positive bacteria transferred via the shell, including *Bacillus*, *Streptococci* and *Enterococci*, which have a higher heat resistance than *Salmonella* and have been described in a number of publications [PRO 07, BAR 07, MIL 10, JAN 11, TEC 14, TEC 15, FOE 87, HID 08, SHA 70, PAY 79]. The storage stages of the pasteurized egg products cause an additional selection pressure on this flora: this is selection by cold temperatures. It allows microorganisms that have resisted pasteurization and are particularly suitable for growth at low temperatures to develop, particularly in the event of a breakage, even if moderate, of the cold chain. Thus, the equilibrium of the shell egg flora is modified by industrial processes. We move from an environment prone to the development of *Salmonella* to an environment prone to the development of a flora that is uncompetitive in shell eggs but which dominates in the egg product and is the cause of spoilage events that must be controlled. Here we provide a description of the main characteristics of this spoiling flora specific to egg products.

2.3.1. Bacillus Cereus sensu lato

2.3.1.1. *Overview*

Bacteria of the *B. cereus* group, also referred to as *B. cereus sensu lato*, are ubiquitous spore-forming, facultative aero-anaerobic Gram-positive bacteria. They possess oval, non-deforming spores in a terminal to sub-terminal position. The group is composed of seven "species": *Bacillus anthracis*, *Bacillus cereus sensu stricto*, *Bacillus mycoides*, *Bacillus pseudomycoides*, *Bacillus thuringiensis*, *Bacillus weihenstephanensis* and *Bacillus cytotoxicus*. This classification is not based on genetic criteria, but rather on subjective practical considerations such as virulence, with regard to *B. anthracis*, *B. cereus sensu stricto*, *B. cytotoxicus* and *B. thuringiensis*, the physiology, with regard to *B. weihenstephanensis*, which can develop at temperatures less than 7°C and cannot grow at temperatures of 43°C or above, and morphology with regard to *B. mycoides* and *B. pseudomycoides*, which form rhizoid colonies on agar medium.

In this group, the species *B. anthracis* is the causative agent of anthrax and *B. thuringiensis* produces an entomopathogenic crystal that can be used as a biopesticide. *B. cereus sensu stricto* is known to cause foodborne illnesses due to the production of two types of toxins. The cereulide or emetic toxin develops in the food prior to ingestion and is the cause of emetic syndrome [AGA 94]. The main manifestations of this syndrome last between 6 and 24 h and are found within the first half hour and up to 6 hours after ingestion of the contaminated food [EML 04]. The associated symptoms are similar to those caused by staphylococcal enterotoxins, proteins preformed by *S. aureus* in food. Several other toxins, the three-component BL hemolysin complex (Hbl B, Hbl L1 and Hbl L2), the non-hemolytic enterotoxin, also composed of 3 subunits (NheA, NheB and NheC) and K cytotoxins (Cyt K1 and Cyt K2), are responsible for diarrheal syndrome [LUN 97], which can be confused with that caused by *Clostridium perfringens*. The toxin BceT has never been involved in episodes of foodborne illness despite its structure classifying it as an enterotoxin [AGA 94]. Diarrheal syndrome is the result of the ingestion of vegetative cells or spores in sufficient quantities, with these toxins being produced by the small intestine. The incubation time is therefore longer than for the emetic syndrome, and initial symptoms are generally expressed 6 h after ingestion, for a period of between 12 and 24 h [GRA 97, CLA 07]. These two syndromes require a significant population of between 5 and 8 log cfu per gram of food. The species *B. weihenstephanensis*, *B. mycoides* and *B. pseudomycoides* have not been described as agents causing foodborne illnesses although they do have a toxigenic potential [PRÜ 99a, STE 02]. It should also be noted that certain strains of *B. thuringiensis*, *B. mycoides*, *B. pseudomycoides* and *B. weihenstephanensis* also contain the hblA gene and can produce the Hbl enterotoxin and cause foodborne illnesses [PRÜ 99a]. Bacteria in this group also produce many virulence factors, such as hemolysins, phospholipases, proteases and collagenases that play a role in infection [MIC 03, SCH 05] and can cause severe cases of septicemia and other infections [TUR 83].

The classification that has just been presented, based on virulence, physiology or morphology, does not, however, correspond to genetic divergence criteria. Several publications [GUI 03, GUI 08, HIL 04] highlight the phylogenetic relationships between the 425 strains representing the six "species" of the group, by the analysis of their fAFLP profiles, the sequences of the *panC* housekeeping gene and by the search for psychrotrophic signatures for the *cspA* and *rDNA16S* gene sequences. These signatures seem to have played a key role in the molecular mechanisms used by these bacteria in adapting to low temperatures. According to [MAY 96], the *cspA* gene is believed to be the most cold-shock-induced homolog in these bacteria. The search for a specific signature on this gene has made it possible to identify the species *B. weihenstephanensis* according to the criteria of [FRA 98]. Moreover, [VON 98] showed that the sequence of the *16S rDNA* gene was discriminant, with the signature rDNA16S-2p characterizing psychrotrophic species and the signature rDNA16S-1m characterizing mesophilic species. [PRÜ 99b] described other signatures (*rrs* signatures) on the same genes.

Even if the use of the species name is still current, for phenotypic criteria – particularly those related to the high virulence of the species *B. anthracis* – a new classification into seven phylogenetic groups can cluster strains of different "species" sharing the same growth temperature range into a single group. This structuring into different phylogenetic groups associated with a growth temperature range indicates that the adaptation of *B. cereus* group bacteria to cold temperatures is an important phenomenon in their evolution [GUI 08]. According to these authors, *B. cereus* group bacteria are thought to have originated from a thermophilic ancestor (which would be represented by group VII) that had evolved into psychrotrophy for adaptation to new ecological niches. The coexistence of two psychrotrophic groups (groups II and VI) seems to indicate parallel evolutions in response to the same constraint. Group VI strains are most strongly identified by their psychrotrophic trait, whereas group II strains appear to have differentiated from mesophilic strains.

2.3.1.2. Bacillus cereus sensu lato *in the egg products industry*

Unlike *Salmonella*, *B. cereus* group bacteria have not been studied for their ability to penetrate eggs. However, they were shown to be unable to grow in egg white, making them unlikely to pose a risk in shell eggs [BAR 14]. The risk is under control as their concentration is low on the shell just after laying, in a breeding environment where hygiene conditions are well controlled [WOO 88, KON 13].

Moreover, despite the strong involvement of *Salmonella* in episodes of collective foodborne outbreaks related to egg consumption, *B. cereus* group bacteria have never been implicated in such issues. In France, between 2006 and 2010, the main food vectors (53%) of *B. cereus* group bacteria corresponded to products based on starchy foods (rice, pasta, semolina, wheat, bulgur), with cooked vegetables, including mashed potatoes (13%), in second place [CAD 12]. However, it should be noted that in 2011, toxins produced by *B. cereus* group bacteria were involved in two cases of foodborne outbreaks potentially related to the consumption of eggs or egg products [EFS 14a, EFS 14b] . The actual involvement of eggs or egg products in these infections is still unknown. Indeed, egg products are intermediate products used in many industrial foods and they are also widely used in mass and commercial catering. These places account for 87% of the French cases of foodborne outbreaks caused by *B. cereus sensu lato*, according to the study of [CAD 12], with 49% occurring in schools. The health risk seems therefore negligible in the egg product sector. In addition, foodborne outbreaks caused by this bacterial group are generally benign, rapidly resolving and require a large population of bacteria. However, as described in the pasteurized milk industry [VIT 16, MAC 17, FLI 17, DOL 17], psychrotrophic strains of this group are involved in the spoilage of pasteurized egg products, due to their sporulation properties, their ability to form biofilms and the expression of spoiling enzyme activities, even at low temperatures [BAR 07, JAN 11, TEC 14, TEC 15].

2.3.2. Streptococcus *and* Enterococcus

2.3.2.1. *General overview*

The genus *Streptococcus* is composed of Gram-positive, catalase-negative, non-motile cocci, organized in pairs or in chains and whose respiratory type is predominantly anaerobic. All strains have complex nutritional requirements and lactic acid is the main product of their carbohydrate catabolism. One of their classification criteria is based on their hemolysis ability. Thus, β-hemolytic streptococci (producing a clear hydrolysis zone immediately around the colony on blood agar) can be distinguished from α-hemolytic streptococci (producing a smaller greenish zone with respect to the zone produced by the β-hemolytic type on blood agar) and non-hemolytics treptococci. Another classification is based on the antigenic properties of a parietal compound, C-polysaccharide [LAN 33]. The β-hemolytic streptococci have thus been divided into several groups ranging from A to T. Human pathogens are mainly in groups A to G. However, this classification is not sufficient to differentiate all the streptococci given that many cocci are non-hemolytic or α-hemolytic and do not possess the antigens targeted by Lancefield's differentiation method (1933) [LAN 33]. The latter is still in force in the medical field to identify certain pathogenic β-hemolytic strains, such as the species *Streptococcus pyogenes* and *Streptococcus agalactiae*, but it does not take into account other taxonomic markers, in particular genetic markers. It also does not disclose the evolutionary links between the different streptococci strains. Since Sherman's research in 1937 [SHE 37], four groups of streptococci have been identified, based on their hemolysis capacity, their antigenic profile, and other phenotypic characteristics (fermentation and tolerance tests). The following groups were identified: pyogenic, viridans, lactic acid and *Enterococcus*. The group of pyogenic streptococci includes the β-hemolytic streptococci, which are pathogenic to animals, including to humans. The viridans group includes non-β-hemolytic and species intolerant to alkaline pH and high salt concentrations and that are unable to grow at 10°C. This group includes commensal species of the upper respiratory tract. The lactic acid group includes strains that were first identified in dairy environments. They are not pathogenic, non-β-hemolytic, are able to grow at 10°C but not at 45°C and are sensitive to an NaCl concentration of 6.5%. The *Enterococcus* group corresponds to α-hemolytic streptococci, although it also includes some β-hemolytic strains. The other characteristics described for this group are the ability to grow at alkaline pH, in the presence of high salt concentrations and over a wide temperature range (10 to 45°C). Also included in this group are commensal species of the digestive tract.

With advances in molecular biology, the determination of the percentage of G+C, the sequencing of ribosomal RNA and DNA-DNA hybridization, streptococci were subdivided into three genera in the 1980s: the genera *Streptococcus*, *Enterococcus* and *Lactococcus*. Changes in classification still occur today [SCH 84, SCH 87]. The genera *Enterococcus* and *Streptococcus* belong to a heterogeneous Gram-positive group with a large number of characters common to the genera *Lactococcus* and *Vagococcus*, and are also classified into the lactic acid group [STI 97, AGU 12]. They become part of the lactic flora, provided that they are of food origin [GIR 02]. Food lactic acid bacteria now include species of the genera *Carnobacterium*, *Enterococcus*, *Lactobacillus*, *Lactococcus*, *Leuconostoc*, *Oenococcus*, *Pediococcus*, *Streptococcus*, *Tetragenococcus*, *Vagococcus* and *Weissella*. The phylogenetic system of the genera *Enterococcus* and *Streptococcus* is therefore not fully understood and is subject to further reclassifications [GIR 02].

2.3.2.2. *The genus* Streptococcus

The genus *Streptococcus* includes about 100 human and animal commensal or pathogenic species. These bacteria are non-motile, Gram-positive, occurring in pairs or in variable length chains and lack a cytochrome. Some species also have a capsule. These bacteria have a low percentage of G+C, and their classification is constantly evolving. In the field of medical bacteriology, the latter is based on their hemolytic character and the fact that they belong to one of the Lancefield serological groups. They are currently classified as “sets” and “subsets”, which has no real taxonomic basis but is used by the medical community. It is possible to distinguish between pyogenic streptococci, group D streptococci, oral streptococci and so-called “unclassified” streptococci. Many species of the genus are pathogenic to humans. *Streptococcus faecalis*, a Lancefield group D *Streptococcus* resistant to penicillin, is a commensal digestive tract bacterium that can cause urinary tract infections and endocarditis. *Streptococcus pneumoniae* is responsible for pneumonia. *Streptococcus pyogenes*, which is β-hemolytic and of the Lancefield group A, is responsible for infections of the respiratory and urinary tracts, and of tonsils or skin. Some other β-hemolytic strains are involved in cases of rheumatic fever or glomerulonephritis. *Streptococcus viridians* is a commensal bacterium of the oral sphere that can cause endocarditis, especially when its route of entry is via the blood.

Since the classification of the genera *Streptococcus*, *Enterococcus* and *Lactococcus* in the 1980s [SCH 84, SCH 87], species of the genus *Streptococcus* include pathogenic and oral sphere streptococci (formerly Sherman *viridians* group) [STI 97]. The exception is the species *Streptococcus thermophilus* which is widely used as a fermenting agent in yoghurt production. Based on DNA–DNA hybridization results, this species belongs to the same cluster as the species *Streptococcus salivarius* and *Streptococcus vestibularis* but it is still considered as a separate species due to its use as a fermenting agent. It is therefore the only fermenting agent classed in this genus [STI 97].

The species *Streptococcus avium*, *Streptococcus durans*, *Streptococcus faecalis*, *Streptococcus faecium* and *Streptococcus gallinarum* were reclassified into the genus Enterococcus and are now known as *Enterococcus avium*, etc. The remainder of this document will be devoted only to the genus *Enterococcus*, including the species *E. faecium* and *E. faecalis*, formerly *S. faecium* and *S. faecalis* respectively, [SCH 84] which are the most likely to contaminate pasteurized egg products [TEC 15].

2.3.2.3. *The genus* Enterococcus

The genus *Enterococcus* is composed of about 30 species predominantly of Lancefield group D. These bacteria are ubiquitous. They are found in the soil, in water, on plants, vegetables, insects [DEI 63] and in the gut of humans and animals [FOU 06]. The species *E. faecalis* and *E. faecium* are the most common in food, followed by *E. casseliflavus, E. hirae, E. gallinarum* and *E. avium* [PRO 07, CAM 06, STI 78, BEL 85, KNU 93, PET 03, KLE 03, TOM 08, AL 10, AND 16]. The species *E. faecalis* and *E. faecium* are also of clinical concern because of their ability to develop antibiotic resistance [ARI 12].

As lactic acid bacteria, some species of the genus *Enterococcus* are used in a number of processes as starters improving the sensory characteristics of foods (taste and smell), as probiotics and/or as conservatives. The species *E. faecium, E. faecalis* and *E. durans*, and to a lesser extent, *E. casseliflavus*, are associated with fermented milk products [FRA 99, GIR 03, BUR 07, DOS 15]. This genus is also identified in certain meat products (sausages, traditional meat products, chickens, pig carcasses) [SAB 04, BAR 10, DEM 17, AYM 06], seafood products [ALB 10], and certain vegetables such as olives [BEN 04, FER 83], sorghum [HAM 97, YOU 05] and soya [MUL 90, YOO 08, DE 05, TOD 05, ZEN 05].

The genus *Enterococcus* can also be used for its probiotic properties (prevention of diarrhea) associated or not with antibiotics [BEL 80, BUY 96], its cholesterol-lowering action [BUY 96], and its antimicrobial activities mainly related to the production of bacteriocins [MAR 04, TOD 09].

Nevertheless, these practices are now controversial, as some strains, particularly *E. faecalis* and *E. faecium*, are implicated in nosocomial diseases or foodborne diseases [AGU 12, FIS 09]. The attention given to *Enterococcus* in public health is due to its potential enteric habitat and, in connection with this, the fact that *E. faecalis* is used as an indicator of fecal contamination [STI 89]. The rationale for choosing this species as an indicator of fecal contamination is based on the fact that these bacteria are able to survive in an extra-enteric environment, are heat-resistant and can become dominant in foods that have undergone heat treatment [STI 97].

As commensal flora of animal digestive tracts, the probability of meat products becoming contaminated by this bacterial genus is high. The most common species are also *E. faecium* and *E. faecalis* [AGU 12] as well as *E. hirae* and *E. durans* at a lower occurrence [FRA 04]. *E. faecium* and *E. mundtii* have also been detected in fish or seafood products [CAM 06, VAL 10]. *Enterococcus* is part of the commensal flora of the digestive tract of chicken, and contamination of carcasses by fecal bacteria is more common than in other meats, possibly due to less stringent hygienic conditions in the chicken industry than in pork and beef production. According to [BOR 16], the prevalence of enterococci-contaminated chicken is generally high, although it depends on sample types, isolation methods and slaughter hygiene conditions. Ninety-six percent of raw meat samples were shown to be contaminated with *E. faecalis* and *E. faecium*, with bacterial concentrations of less than 2 log cfu per gram [GIR 03, MIR 07].

The most prevalent species in human infections is *E. faecalis* (80–90% of infections by this genus), followed by *E. faecium* (5–10% of infections by this genus) [KAY 03, SÁN 07]. These two species now appear to be the third most common bacteria causing bacteremia in Europe and the United States of America (USA), after *E. coli and S. aureus* [DEK 13]. Human infections due to *E. durans, E. avium, E. gallinarum, E. irae, E. casseliflavus, E. raffinosus*, and *E. mundtii* have also been reported [JET 94].

The genetic plasticity of the genus *Enterococcus*, particularly the ability to acquire mobile genetic elements, allows these bacteria to adapt to many ecosystems and to be a vector for (1) broad-spectrum antimicrobial resistance factors [INO 06, LEB 06], including for strains carried by chicken [BOR 16] and (2) virulence factors, the best-described being the aggregation substance [JET 94, CHO 93, EAT 01], β-hemolysin orcytolysin [LEB 06, CHO 93, HUY 91] and hydrolases [FIS 09]. Enterococci transmitted by contaminated food, including chicken, can colonize the human intestine, mainly the colon, where they account for a small proportion (<1%) of the cultivable microbiota but can be pathogenic and cause antibiotic resistance gene transfer to native flora [LEB 14]. Enterococci are inherently resistant to different classes of antimicrobials; there is great diversity in antimicrobial resistance profiles and, to a lesser extent, in the genetic basis for the acquisition of these resistances among *E. faecium* and *E. faecalis* [ARI 12]. In any case, the horizontal transfer of resistance genes is the greatest risk posed by enterococci from chicken. Gentamicin resistance genes carried by *E. faecium* and *E. faecalis*, and quinupristin/dalfopristin resistance genes carried by *E. faecium* are currently of greatest concern, although differences are observed from one geographical area to another [BOR 16].

The genus *Enterococcus* may also be responsible for the formation of biogenic amines in food [KAR 05]. The most sensitive food products are fermented products in which *Enterococcus* is used, because of the extent of proteolysis and acidic conditions prevalent in these products.

All the arguments presented show the opportunistic pathogenicity of enterococci and the risk that they pose as vectors of antimicrobial resistance. It will be easily understood why this genus does not have access to the Generally Recognized as Safe (GRAS) and Qualified Presumption of Safety (QPS) statuses. Each strain used as a fermenting agent or as a probiotic is first subjected to an evaluation that must attest to its safety, in particular by identifying factors that could promote gene transfer and the emergence of opportunistic virulence traits [AGU 12].

2.3.2.4. *The genus* Enterococcus *in the egg product industry*

Bacteria of the genus *Enterococcus* are among the most thermotolerant non-sporulating bacteria [PER 82, MAG 88]. They are able to adapt to many substrates and growth conditions (low and high temperatures, extreme conditions of pH and salinity) [MOR 06]. This genus can potentially grow in foods with a_w values greater than 0.92, in a temperature range of 5°C to 50°C, with an optimum temperature of 35°C, and pH ranging from 4.6 to 9.9, with an optimum around 7.5 [FIS 09].

Due to their ubiquitous nature and their heat resistance, these bacteria are contaminants frequently encountered in some heat-treated products [MOR 06]. The genus *Enterococcus* has been particularly highlighted during spoilage of milk, cheese (cheddar), fish, or meat products such as ham or some pâtés [MAG 88, MOR 06, MCK 84,

JEN 75, MAG 86, VAS 08]. Knowing that this genus is part of the commensal flora of the digestive tract of chicken [BOR 16], it is not surprising to find it on the shell and, subsequently, in the egg product. In the study by [MIL 10] on egg products, lactic acid bacteria – represented by the species *E. faecalis* and *E. faecieum* – are the dominant flora appearing during refrigerated storage of pasteurized whole egg and are responsible for its spoilage. Of the 45 isolated strains, 67% belong to the species *E. faecium* and 27% to the species *E. faecalis*. All these strains withstand a pasteurization treatment of 60°C for 30 min, and even 80°C for 3 min for some strains of *E. faecium*. In one study [TEC 15], samples of raw egg white were analyzed and 39% were found to be contaminated with *Enterococcus*. There was a higher prevalence in the contaminated pasteurized samples (93%), suggesting that these bacteria had resisted the heat stabilisation process. It should be noted, however, that the microbial contamination levels reported in this study are very low, with less than 2 and 1 cfu/mL mesophilic aero-anaerobic bacteria in raw egg white and in pasteurized egg white, respectively. It is important to bear in mind that egg white is poorly suited to harbor microorganisms and permit their growth.

2.4. Bacterial metabolism that can affect whole egg

2.4.1. *The whole egg as an optimal culture medium*

On average, the liquid contents of an egg equate to 32% egg yolk and 68% egg white, consisting of approximately 73.6% water, 12.8% protein, 11.8% fat, 1% carbohydrate and 0.8% minerals [GUE 10, YAM 10, MAJ 17, HOR 17]. It provides all the nutrients necessary for bacterial growth. If we take the example of *B. cereus* group bacteria, their growth capacity in whole egg exceeds even that observed in a rich laboratory medium such as brain heart infusion (BHI) to which yeast extract is added (BHI-YE) [TEC 14].

We will focus on the most representative classes of molecules – such as proteins, lipids and carbohydrates – but also minerals and vitamins, which are essential nutrients for bacterial growth and expression of spoiling activities, particularly those involving proteases, phospholipases and lipases, known to be at the origin of spoilage of various foods, milk being the best-described example [MAC 17].

The term proteolytic enzyme includes all hydrolases that act on proteins or hydrolysates derived from these proteins. Other terms refer to them in the literature as "peptide-bond hydrolase", peptidase or protease. With regard to proteolytic activities of bacteria when spoiling pasteurized foods, milk is one of the most commonly described examples [MAC 17, LED 09]. These activities, usually heat-resistant [GRI 81, COG 77, SAO 16], are manifested by a hydrolysis of casein causing the coagulation of milk – the release of (1) hydrophobic peptides which cause the development of a bitterness and (2) amino acids that cause an increase in pH. [GRA 90, DAL 93, DAI 98, BAI 05, BEN 04].

Biogenic amines are low molecular weight organic bases with aliphatic, aromatic or heterocyclic structures. They are formed by the decarboxylation of amino acids due to the action of microbial enzymes and can also result from the hydrolysis of nitrogen compounds or the amination of aldehydes or ketones [ASK 86]. Histamine is mainly produced by histidine decarboxylation catalyzed by histidine decarboxylases, tyramine by tyrosine decarboxylation catalyzed by tyrosine decarboxylases and putrescine by the decarboxylation of ornithine (arginine derivative) catalyzed by ornithine decarboxylases or by the deimination of agmatine catalyzed by agmatine deiminases. Biogenic amines may cause different problems depending on the ingested dose and the state of health of the affected individual, ranging from headaches or allergic-type symptoms to hypertensive crisis or food poisoning [SHA 96].

Lipases are carboxylesterases that belong to a class of hydrolytic enzymes with a broad spectrum of biological activities, including catalyzing the conversion of triglycerides to diglycerides, monoglycerides, fatty acids, and glycerol [JAE 94, CHE 03]. These enzymes are widely distributed in the animal, plant and microbial kingdoms. Microbial lipases have the peculiarity of catalyzing reactions as varied as hydrolysis, interesterification, esterification, alcoholysis and acidolysis [BRO 84, VER 97]. Lipases exhibit different types of selectivities: typo-selectivity (specificity with respect to certain fatty acids), regio-selectivity (specificity of position of the fatty acid on the glycerol skeleton), enantio-selectivity (specificity towards one enantiomer with respect to another in a racemic mixture or in the case of a chiral molecule). True lipases or triacylglycerol acylhydrolases (EC 3.1.1.3) which act at oil–water interfaces [VER 97] differ from esterases or carboxylases (EC 3.1.1.1), active in aqueous solutions, by the fact that they hydrolyze esters of glycerol-bearing long chain ($C > 10$), or short chain ($C < 10$) fatty acids, respectively.

Lipase activities are also described in milk where they cause rancid, dirty, fruity, soapy odors and/or bitter tastes, making the food unfit for consumption [COU 82, DEE 06]. In milk, the hydrolysis of just 1–2% of the triglycerides into short-chain fatty acids causes odors of rancidity [INT 96]. Their action on milk causes the release of short chain (C4:0 to C8:0), medium chain (C10:0 and C12:0) and long chain (C14:0 to C18:0) fatty acids. Short-chain fatty acids have a marked odor and may be responsible for rancid, bitter, "butyric", dirty, astringent or "lipase" odors [DEE 11]. Medium-chain fatty acids give a soapy taste to milk [CHE 03]. Long-chain fatty acids have less effect on the flavor of milk. In addition, an oxidized flavor can come from the oxidation of unsaturated fatty acids to aldehydes and ketones [CHE 03]. Mono- and diglycerides, which are also the hydrolysis products of lipases, can modify the surface properties of milk lipids [DEE 11].

Regarding the involvement of phospholipase activities in food spoilage, we also address the research carried out on dairy products, which show that the *B. cereus* phospholipase C causes the degradation of the membranes of fat globules causing the formation of lumps in the cream [FRA 07]. The destabilization of milk fat is a phenomenon known as "bitty cream". Under the action of phospholipases, this phenomenon is manifested by the appearance of aggregates on the surface, due to the destabilization of fat globules. It also makes the milk more likely to be hydrolyzed by lipases [COU 82, SHA 94, MEE 91]. As for proteases, certain lipases and phospholipases can persist and give rise to organoleptic defects after heat treatment, while the microorganisms responsible are destroyed [GRI 81, COG 77, SAD 16, COU 82, CHE 04].

While few studies have been devoted to the spoilage of egg products, we will see how the richness of the egg in metabolizable substrates and the abundance of hydrolases described in the *B. cereus* group and the genus *Enterococcus* meet the conditions leading to spoilage issues, as already described for other types of foods.

2.4.1.1. *Whole egg proteins [GUE 10, HOR 17]*

Whole egg comprises on average 12.8% proteins, of which 4% are from the egg yolk and 8.7% are from the egg white. The average amino acid composition of the egg white broadly reflects that of the whole egg (Table 2.1).

Amino acid	% in egg white (m/m)	Amino acid	% in whole egg (m/m)
Glu	1.36	Glu	1.54
Asp	1.06	Asp	1.18
Leu	0.88	Leu	1.04
Val	0.72	Ser	0.91
Ser	0.71	Lys	0.82
Lys	0.66	Val	0.79
Ala	0.64	Arg	0.77
Phe	0.61	Ala	0.69
Arg	0.60	Ile	0.66
Ile	0.56	Phe	0.64
Thr	0.47	Thr	0.59
Tyr	0.41	Tyr	0.51
Pro	0.40	Pro	0.48
Met	0.39	Gly	0.40
Gly	0.36	Met	0.39
Cys	0.28	His	0.29
His	0.24	Cys	0.28
Trp	0.17	Trp	0.19

Table 2.1. *Amino acid composition in egg white and whole egg [GER 07]*

The majority of egg yolk proteins are lipid-associated forming high-density lipoproteins (HDLs, representing 36% of egg yolk proteins) and low density lipoproteins (LDLs, representing 22% of egg yolk proteins). In LDLs, we find apo-vitelline 1 and apolipoprotein B. In HDLs, we find lipovitellins 1 and 2. These proteins are highly hydrophobic (30 to 40% hydrophobic amino acids) and can be glycosylated by hexoses, hexosamines and sialic acids and/or phosphorylated on serine residues. Other free-form yolk proteins are globular proteins, livetins (30%), phosphoglycoprotein, phosvitin (about 10%) – which can form a complex with HDLs – and minor proteins (about 2%).

Egg white proteins have variable molecular weights, ranging from 12.7 to 8,000 kDa and isoelectric points (pI) ranging from 4 to 11. Some of them are present in high concentrations, such as ovomucin, which alone accounts for 54% of egg white proteins; others occur only in trace amounts. Most of these proteins are glycosylated, mainly at asparagine (N-glycosylations), serine or threonine (O-glycoslysations) residues. Carbohydrate moieties, which can account for more than 60% of the weight of the protein, include various sugars, including D-glucose, D-mannose, D-galcatose, fucose, 2-amino-2-deoxy-D-glucose, N-acetylglucosamine, N-acetylgalactosamine, sialic acid, and N-acetylneuraminic acid. Some proteins are also phosphorylated on certain serine residues included in the Ser-X-Glu sequence.

At present, there are at least 12 families of different proteins in egg white:

1) serpins, represented by the phosphoglycoprotein ovalbumin (54% of egg white proteins);

2) transferrins, represented by the glycoprotein ovotransferrin (also known as conalbumin) (12–13% of proteins);

3) antiproteases, represented by the glycoproteins ovomucoid (11% of white proteins), ovo-inhibitor (0.1–1.5% of proteins) and ovostatin (0.5 % of proteins) and by the phosphorylable protein cystatin (0.05% of proteins);

4) glycoside hydrolase family 22, represented by lysozyme (3.4–3.5% of proteins);

5) mucins, represented by the sulfated glycoprotein ovomucin (1.5–3.5% of proteins);

6) lipocalines, represented by the glycoprotein ovoglycoprotein (0.5–1% of proteins);

7) folate receptors, represented by the phosphoglycoprotein riboflavin-binding protein (RBP) (0.8% of proteins);

8) the avidin/streptavidin family, represented by the glycoprotein avidin (0.05% of proteins);

9) bactericidal-permeability-increasing proteins (BPI), represented by the protein TENP (transiently expressed in neural precursors) (% unknown);

10) clusterins, represented by the glycoprotein clusterin (% unknown);

11) the urokinase plasminogen activator receptor (uPAR)/CD59/Ly6/snake neurotoxin superfamily, represented by the protein HEP21 (% unknown);

12) proteases, represented by glutamyl aminopeptidase, methionine aminopeptidase (MAP) and matrix metalloprotease-2 (MMP-2) (in trace amounts).

2.4.1.2. *Whole egg lipids [GUE 10, HOR 17]*

Since egg white only contains lipids in trace amounts, the lipids of whole egg come exclusively from the yolk. Lipids, which account for nearly 31% of egg yolk, consist of 62% triglycerides, 29% phospholipids, 5% cholesterol, less than 1% free fatty acids and traces of other lipids, including carotenoids responsible for its color. These lipids are distributed exclusively as lipoproteins (HDLs and LDLs). LDLs consist of nearly 90% lipids, which are divided into neutral lipids (triglycerides and cholesterol esters) surrounded by a monolayer of phospholipids, proteins and cholesterol, in contact with the aqueous phase. HDLs are composed of 20% lipids, of which 65% are phospholipids and 35% are neutral lipids. The overall fatty acid composition of egg yolk lipids varies depending on the diet of the hen, particularly the monounsaturated fatty acid (MUFA) and polyunsaturated fatty acid (PUFA) contents. However, the saturated fatty acid content is little affected by the nature of the fatty acids supplied by the diet. These modifications primarily concern the fatty acids of the triglycerides. The overall fatty acid composition of egg yolk is between 30 and 35% for saturated fatty acids (SFA), 40 and 45% for MUFA and 20 and 25% for PUFA. The yolk fatty acids are, in order of abundance, oleic acid (40–45% of total fatty acids), palmitic acid (20–25%), linoleic acid (15–20%), stearic acid (14%); α-linolenic, arachidonic and docosahexaenoic acids occur in smaller proportions (close to 1% for each).

This distribution also corresponds to that found in egg yolk triglycerides. On the glycerol skeleton, we find palmitic acid at *sn*-1, linoleic acid in the *sn*-2 position and stearic acid in the *sn*-3 position. Oleic acid is also distributed between the *sn*-2 and *sn*-3 positions. Figure 2.1 presents the major molecular species representing egg yolk triglycerides; these species account for 69% of the total.

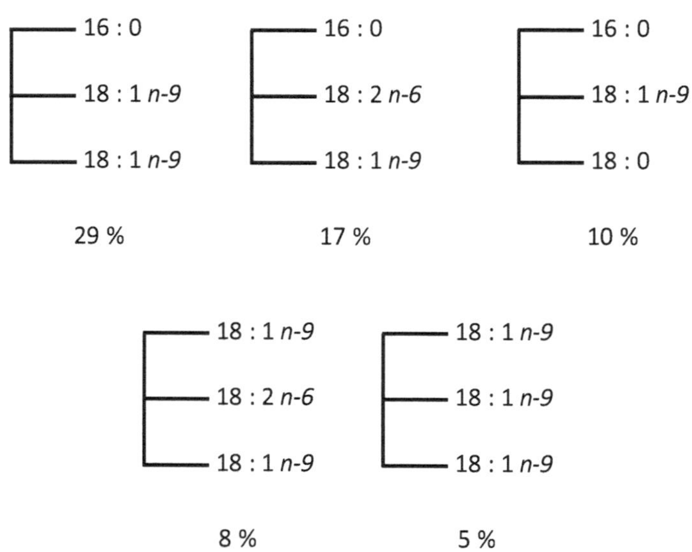

Figure 2.1. *Main molecular species representing egg yolk triglycerides*

Phospholipids are composed of higher proportions of polyene species harboring more than three unsaturations (35–40%) than triglycerides (16%). SFAs and MUFAs represent approximately 45% and 20–25% of phospholipid fatty acids, respectively. As for the class of phospholipids present in egg yolk, phosphatidylcholine (PC, 76%) and phosphatidylethanolamine (PE, 20.5%) are primarily found. Phosphatidylserine (PS), LysoPC and PE, sphingomyelin (SM), phosphatidylinositol (PI) and cardiolipids each do not account for more than 1% of the total phospholipid content.

The fatty acids constituting PC are SFA (44%; 31% palmitic and 12% stearic acids), MUFA (28%; 25% oleic acid) and PUFA (28%; 20% linolenic acid). The SFA are mainly positioned at *sn-1* of PC whereas USFA are at the *sn-2* position. The main molecular species of PC are presented in Figure 2.2.

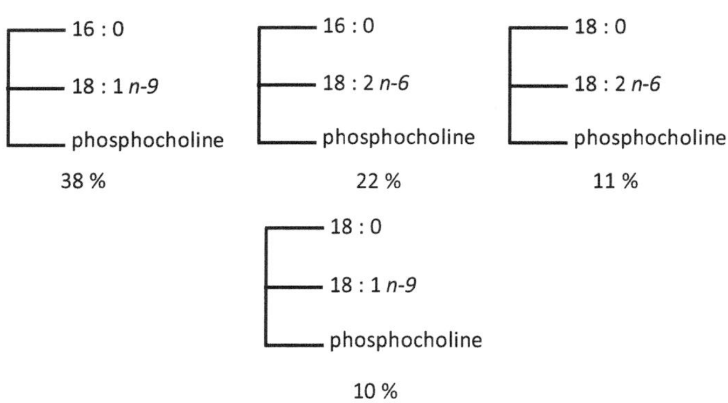

Figure 2.2. *Main molecular species of phosphatidylcholine in egg yolk*

The fatty acids of PE are SFA (43%) – mainly palmitic (16%) and stearic (26%) acids, PUFA (39%) – mainly linolenic and arachidonic acids (14% each), and MUFA (18%) – mainly oleic acid (16%). It is in this class of lipids that long chain PUFA are preferentially incorporated. The SFA are mainly positioned at *sn-1* of the PE whereas the

USFA are in the *sn-2* position. We can also find alkyls or alkenyls preferentially positioned at *sn-1* but their quantity is low (<3%). The main molecular species of PE are shown in Figure 2.3.

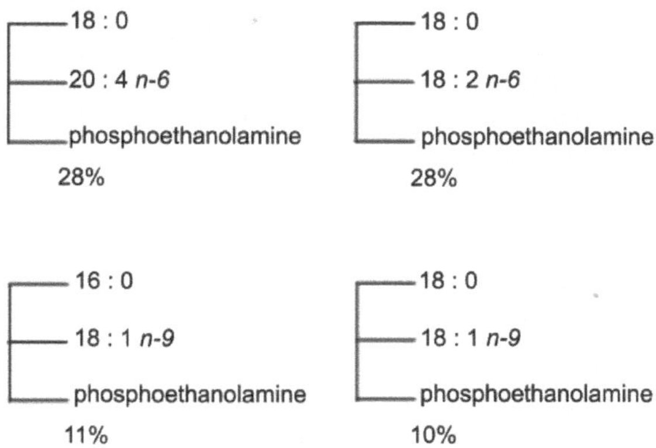

Figure 2.3. *Main molecular species of phosphatidylethanolamine in egg yolk*

Cholesterol accounts for about 5% of egg yolk lipids. It is found mainly in free form (85–90%). Cholesterol esters, which therefore represent 10 to 15% of total egg yolk cholesterol, are found in the LDL core together with triglycerides. These are cholesteryl oleate (35%), cholesteryl palmitate (33%), cholesteryl linoleate (12%) and cholesteryl stearate (11%).

2.4.1.3. *Whole egg carbohydrates [GUE 10, HOR 17]*

Egg yolk carbohydrates, which account for 0.7 to 1% of its total components, are partly free glucose (0.3%) but mainly oligosaccharides associated with proteins by N- or O-glycosidic bonds (glycoproteins) or lipids (glycolipids such as cerebroside). It is the same for egg white carbohydrates, which account for 0.9%, with half represented by free glucose and half by monosaccharides such as galactose and mannose, N-acetyls, amino sugars, uronic and sialic acids, and protein-bound oligosaccharides. These oligosaccharides are linked to proteins by N-glycosidic bonds (ovotransferrin, ovalbumin, ovomucoid and avidin) or O-glycoside bonds (β-ovomucin).

2.4.1.4. *Whole egg minerals [GUE 10, HOR 17]*

The mineral concentrations described in the literature vary because of their dependence on many parameters, including the zootechnical parameters of egg production and more particularly the feeding of hens. Table 2.2 provides the average whole egg mineral composition.

Mineral	mg/100 g
P	193
Cl	172
S	164
K	125
Na	120
Ca	50
Mg	12
Fe	1.7
Zn	1.3
Cu	0.06
Ile	0.05
Mn	0.04
Se	0.025
Cr	~0.0085
Mo	~0.0060

Table 2.2. *Average whole egg mineral composition. Ile: isoleusine*

An egg has a low sodium and calcium content compared to other physiological fluids. Egg yolk has a high phosphorus content, mainly in the form of phospholipids. Almost all the iron and other minerals are present in the egg yolk. Egg white is indeed iron-free and the few traces of iron it contains are chelated with ovotransferrin. In egg yolk, about 90% of the iron is chelated with phosvitin.

2.4.1.5. *Whole egg vitamins [GUE 10, HOR 17]*

The egg white, which contains few vitamins, contains only water-soluble vitamins of group B, and no fat-soluble vitamins. The most abundant vitamins are biotin (vitamin B7), niacin (vitamin B3) and riboflavin (vitamin B2). Most of them are related to egg white proteins. Fat-soluble vitamins (vitamins A, D, E and K) are exclusive to egg yolks. Their contents are variable because they are strongly dependent on the feeding of hens. Table 2.3 shows an average composition of whole egg vitamins.

Water-soluble vitamin	**mg/100g**
Thiamine (B1)	0.09
Riboflavin (B2)	0.33
Niacin (B3)	0.08
Pantothenic acid (B5)	1.52
Pyridoxine (B6)	0.12
Inositol (B7)	10.8
Biotin (B8)	20
Folic acid (B9)	0.06
Cobalamin (B12)	0.88
Fat-soluble vitamin	**UI, mg or µg/100g**
Retinol A, IU	480
Calciferol D, IU	50
Tocopherol E, mg	1.6
Vitamin K, µg	4.25

Table 2.3. *Average composition of whole egg vitamins*

2.4.1.6. *Whole egg color [GUE 10, HOR 17]*

Carotenoids, which account for less than 1% of egg lipids, come exclusively from the chicken's diet and are responsible for the color of the egg yolk, which ranges from very light yellow to dark orange. The main representatives of these lipids are carotenes (β-carotene) and xanthophylls (lutein, cryptoxanthin and zeaxanthin). Lutein and zeaxanthin, however, are the most common carotenoids in egg yolk.

To conclude, the egg, intended to meet the nutritional needs of the embryo during its development, can also, once its natural defenses are overcome – which is the case following the breaking of the shell for the production of food such as pasteurized whole egg – become an ideal growth medium for many microorganisms resistant to the the heat treatments implemented in the egg product industry.

2.4.2. *Metabolism of* B. cereus sensu lato

2.4.2.1. *Central metabolism*

According to [ROS 05], understanding the central metabolism of *B. cereus* group bacteria is complex, due to the great diversity of strains studied, their amino acid requirements for growth and their implementation in often complex environments allowing them to mimic the conditions they encounter in food.

In defined environments, studies by [ROS 05, ZIG 06, OUH 06] and [MOL 07] show that *B. cereus* group bacteria are able to catabolize a wide range of carbohydrates via the Embden-Meyerof pathway and, to a lesser extent, the pentose phosphate route. During fermentation growth under anaerobic conditions (Figure 2.4), pyruvate and acetyl-CoA (acetyl coenzyme A) serve as electron acceptors. Pyruvate is converted to acetyl-CoA by pyruvate from atelyase (PFL) or is reduced to lactate by lactate dehydrogenase (LDH), malate by malate dehydrogenase (MDH), and succinate by succinate dehydrogenase (SDH). Acetate production is influenced by the activity of phosphotransacetylase (PTA) and acetate kinase (ACK). The main fermentation products are therefore acids such as lactic acid, succinic acid, acetic acid, formic acid, ethanol and CO_2. Lactic acid accounts for more than 65% of glucose degradation products in the *B. cereus* strain used in the [ZIG 06] study. Acetic and formic acids are each produced at 15% of the total production. Ethanol and succinate are minor products, representing just 2–3%, respectively, of the total products formed [ROS 05, DUP 06].

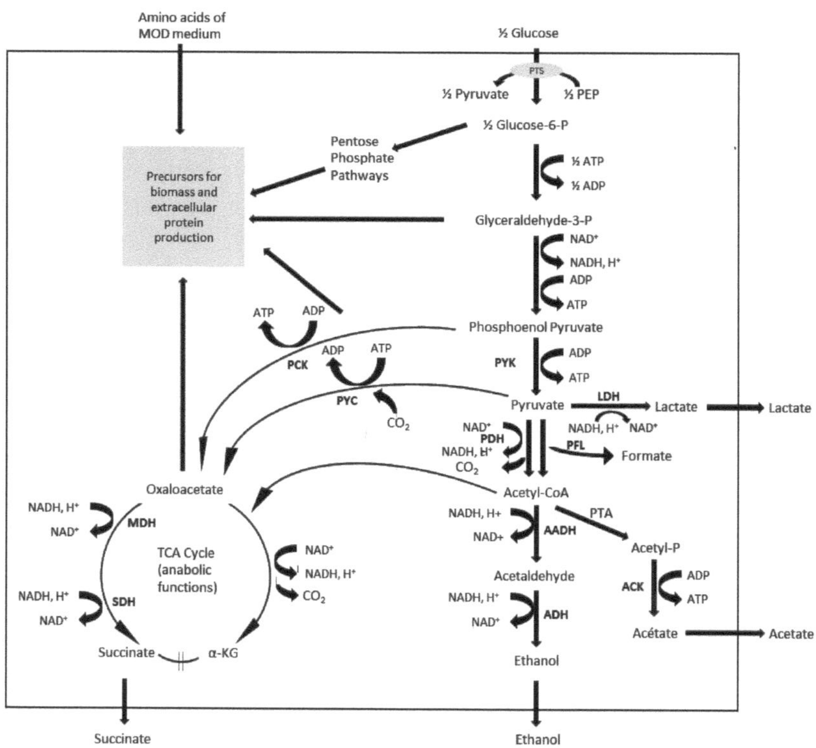

Figure 2.4. *Schematic of anaerobic catabolism of glucose in* B. cereus *[ZIG 06]*

Aerobically, pyruvate is converted to acetyl-CoA by the pyruvate dehydrogenase complex (PDH) and acetate production can occur when the carbon input rate in the Krebs cycle exceeds its oxidation capacity [ROS 05, DUP 06]. However, under anaerobic conditions, *B. cereus* needs carbohydrates for growth (such as glucose, fructose or sucrose) in addition to amino acids [OUH 06, DUP 04, OUH 09]. In the absence of oxygen, the behavior of pathogenic food-borne *B. cereus* group bacteria is extremely dependent on the sugar present in the medium, and any modification of the metabolic activity has consequences on the pathogenicity of the bacteria, in particular their capacity to synthesize enterotoxins [OUH 06, DUP 04, OUH 09].

In *B. cereus*, the expression of most genes encoding metabolic enzymes is modulated by a two-component system, ResDE [DUP 06], a one-component system, the FNR protein [ZIG 07] allowing the expression of virulence genes in low-potential redox environments such as the animal gastrointestinal tract [MOR 04] and catabolite control protein A (CcpA) [VAN 08]. These systems also contribute to the virulence of *B. cereus* [DUP 06, ZIG 07, VAN 08]. Moreover, [LAO 14] have shown that Rex, a regulator of central anaerobic metabolism in *B. subtilis* and other Gram-positive bacteria, also plays a key role in the fermentative and respiratory metabolism of *B.cereus*, in its resistance to reactive oxygen species and in its toxinogenesis.

Sequencing the genome of the strain *B. cereus* ATCC 14579 revealed the complexity of the central metabolic regulation network of *B. cereus* group bacteria. Indeed, more than 200 transcriptional regulators have been highlighted, controlling the identified 5,370 open reading frames (ORFs) [IVA 03, GOH 05]. [GOH 05] have identified about 500 proteins in the extracellular proteome of this strain, which should reflect the importance of protein secretion in pathogenesis in this microorganism. The main proteins identified are collagenases (ColA), metalloproteases (InhA1, InhA2, InhA3, Camelysin, CalY), a protease of the subtilase family (SFP), neutral proteases of the bacillolysin family (NprB and NprP2), a transglutaminase (TGc), aparietal endopeptidase of the family M23/M37 (BC3698), phospholipases (phosphatidylinositol-specific phospholipase C, PI-PLC; phosphatidylcholine-specific phospholipase C, PC-PLC; sphingomyelinase, SMase), achitinase (chitin-binding protein A, ChbA), a 5'-nucleotidase (UshA), hemolysins, enterotoxins and flagellins, all of which have a virulent potential.

2.4.2.2. *Proteolytic activities*

The abundance of genes encoding (1) proteolytic enzymes, (2) peptide and amino acid transporters [IVA 03], and (3) enzymes involved in amino acid catabolism pathways for growth and sporulation [ROS 05, NOL 07, VRI 05a, VRI 05b] reveal the preference of *B. cereus* group bacteria for amino acids. A total of 51 CDS encoding proteases have been identified in the genome of the strain ATCC 14579 [IVA 03]. The extracellular proteome also shows the abundance of peptidases and proteases when the cells are incubated in a protein-rich medium. The hydrolysis activity of proteins and peptides must help cells use exogenous peptides as a source of amino acids [GOH 05, LUO 07]. The enzymes involved in protein metabolism and identified in the proteomic study conducted [LUO 07] on the same strain are presented below (the name of the gene, where known, is written in italics) :

1) Metallopeptidases

i) Class EC:3.4.24.

Collagenase EC, *colA*, EC:3.4.24.3

Bacillolysin, *nprE*, EC:3.4.24.28

O-sialoglycoprotein Endopeptidase, EC:3.4.24.57

Immune inhibitor A precursor, *InhA*, EC:3.4.24.-

Membrane metalloprotease, EC:3.4.24.-

Oligoendopeptidase F, EC:3.4.24.-

Zinc metalloprotease, EC:3.4.24.-

Cell envelope-bound metalloprotease (camelysin), EC:3.4.24.-

ii) Class EC 3.4.99.

Zinc Protease

2) Serine proteases

i) Class EC:3.4.21.

Glutamyl Endopeptidase, EC:3.4.21.19

Prolyl Oligopeptidase, EC:3.4.21.-

Lon-like ATP-dependent protease (Endopeptidase La), EC:3.4.21.-

Thermitase, EC:3.4.21.66

Signal Peptidase I, EC:3.4.21.89

Endopeptidase Clp, EC:3.4.21.92

Minor extracellular protease VPR precursor, EC:3.4.21.-

Protease Do, EC:3.4.21.-

Protease Hhoa, EC:3.4.21.-

i) Class EC:3.4.25.

ATP-dependent protease hslV, EC:3.4.25.2

3) Class EC:3.4.19 exopeptidases

Acylaminoacyl-peptidase, EC:3.4.19.1

Pyroglutamyl-peptidase, EC:3.4.19.3

4) Carboxypeptidases

i) Class EC 3.4.13.

Aminoacyl-histidinedipeptidase, EC:3.4.13.-

Xaa-pro Dipeptidase, EC:3.4.13.9

MembraneDipeptidase, EC:3.4.13.19

ii) Class 3.4.17.

Thermostable Carboxypeptidase 1, EC:3.4.17.19

5) Class EC:3.4.11 aminopeptidases

Leucyl Aminopeptidase, pepA, EC:3.4.11.1

Prolyl Aminopeptidase, EC:3.4.11.5

Xaa-pro Aminopeptidase, EC:3.4.11.9

Methionyl Aminopeptidase, EC:3.4.11.18

X-Pro dipeptidyl-peptidase, EC:3.4.14.11

Aminopeptidase YwaD, EC:3.4.11.6.3.4.11.10

Deblocking aminopeptidase, EC :3.4.11.-

Tripeptide aminopeptidase (Peptidase T), EC:3.4.11.4

According to [CHI 06], the two major proteases secreted by the *B. cereus* group at the end of the stationary phase are Neutral Protease A (NprA) and Immune inhibitor A1 (InhA1). They are metalloproteases, enzymes of 38 to 40 kDa, active at near-neutral pH and requiring calcium for stability and zinc for catalysis [PRI 77]. NprA and InhA1 are recognized as being involved in virulence (toxin activation) [CHU 06, SID 79], bypassing the host's immune defenses [EDL 76, RAM 05, CAD 10] or facilitating the access of bacteria to host deep-tissues by hydrolysing structural proteins [SID 79, MIY 00].

The spoiling proteolytic activities due to the genus *Bacillus* are widely described in the literature [NAB 10,

BAC 99, BAC 01, GRA 04] and also with emphasis on the dairy industry [CHE 03, CHE 04, PRI 77, YAN 85, KAL 92, PIR 96, LIN 00]. According to [NAB 10] and [MAC 17], *Bacillus* spp. show more diverse proteolytic activities than *Pseudomonas,* the genus most implicated in the enzymatic spoilage of dairy products. In addition, many species of *Bacillus* are able to express several types of both extra- and intracellular peptidases. Certain *Bacillus* enzymes are exploited at the industrial level, mainly for their use in detergent powders, as they are adapted to high temperatures because of their particular heat resistance [MAC 94, GUP 02, WAR 83]. However, to our knowledge only the study by [TEC 15] describes the proteolytic activity of *B. cereus* group bacteria isolated from industrial egg products. All strains isolated in this study, involved in the spoilage of a refrigerated dessert made from egg products, the commercial "floating island", appear to develop proteolytic activities. However, the link between the enzymatic activities measured *in vitro* and the physico-chemical changes accompanying the spoilage of the dessert are not described.

2.4.2.3. *Phospholipase activities*

Like other pathogenic bacteria [TIT 98], *B. cereus* group bacteria also produce several types of phospholipase C, which, by hydrolyzing membrane phospholipids, can also facilitate the invasion of host tissues and alter cell signaling [DRO 93, BEE 00, ZEN 05]. Three phospholipase C are described in the *B. cereus* group, phosphatidylinositol phospholipase C (PI-PLC) coded by the *plcA* gene, phosphatidylcholine phospholipase C (PC-PLC) coded by the *plcB* gene (*cerA*, in the operon *cerAB*) and sphingomyelinase (SM-PLC or SMase) coded by the *cerB* gene (in the operon *cerAB*). PC-PLC and SM-PLC can combine to form the AB cerolysin described for its cytolytic activity [GIL 89]. The genes *plcA*, *cerA* and *cerB* are widely distributed in the *B. cereus* group [KIM 08, MAR 09]. The gene *plcA* is not grouped with the other two and shares no similarities with them [KUP 89].

PC, PS and PE are the substrates of PC-PLC [MAR 00]. The action of phospholipase C thus leads to the release of a diglyceride and a compound containing a phosphate group (phospho-choline, phosphoserine or phosphoethanolamine). PI-PLC is specific for phosphatidylinositol and also glycolipids derived from it [HON 98]. SM-PLC specifically cleaves sphingolipids. The phospholipase C activity of the *B. cereus* group is one of the main biochemical properties used for its detection using official methods [PIR 00]. The richness of egg yolk in lecithin makes egg products prone to spoilage by the activity of this enzyme. The activity of this enzyme can be seen in egg yolk-based emulsions or agar plates [KUS 57, HER 95, FRI 08]. In an egg yolk emulsion, lecithinase produces a whitening of the medium and forms a lipid layer on the surface [MAC 41]. It reduces the binding power of egg yolk and causes a marked change in its colloidal state [TEC 14]. On an agar plate containing 5.0–5.5% egg yolk, colonies producing lecithinase develop an opaque halo [MCG 48]. On the other hand, the standard isolation and enumeration methods of *B. cereus* group bacteria are carried out on isolation media based on egg yolk (e.g. polymyxin-egg yolk-mannitol-bromothymol blue (PEMBA) or mannitol-egg yolk-polymyxin (MYP)) [SCH 00].

To our knowledge, only the study by [TEC 15] describes the phospholipase activity of *B. cereus* group bacteria derived from industrial egg products involved in the spoilage of a refrigerated dessert made from egg products, the commercial "floating island". As for proteases, any link could be established between the enzymatic activities measured *in vitro* and the physico-chemical changes observed overthe course of spoilage.

2.4.2.4. *Lipase activities*

The genus *Bacillus* is also well described as a lipase producer [MAC 17, GUN 11]. The published studies focus on the conditions allowing the expression of lipase or esterase activities in the ecological niche from which the bacterium is derived [CHE 04, KAL 92, PIR 96, LIN 00, KIM 02, GHA 00, DAR 94, RUI 05, ERT 07, CHE 07]. The lipase activities involved in food spoilage are mainly described in milk and milk products. While *Pseudomonas* is the main representative of Gram-negative bacteria expressing these activities, the genus *Bacillus* is the most implicated among Gram-positive bacteria [DEC 17]. Most *Bacillus* lipases from dairy environments are active at an optimum temperature above 60°C. These enzymes are also able to withstand heat treatments, including UHT treatments. The thermophilic species *Geobacillus thermocatenulatus* and *Geobacillus stearothermophilus* are the best examples [VIT 16, MAC 17, CHE 03, CHE 04]. Among the studies on the exploitation of microbial lipase activities, research is focused on the genus *Bacillus*, which can produce high concentrations of extracellular lipases in a wide range of pH and temperatures [VAS 16]. *Bacillus* lipases are thus exploited for industrial applications, including for the production of detergents due to their high-temperature fat hydrolysis properties [BAL 16, SAR 17]. This is the case for a lipase produced by a *B. cereus* group bacterium [GUN 11].

Regarding the involvement of lipase activities in the spoilage of egg products, only the work of [MIL 10] and [TEC 15] report it. [MIL 10] show that approximately 87% of strains isolated from liquid whole eggs show lipolytic activities. In the study by [TEC 15], this activity is highlighted for 60%, 50%, 85.7%, 66.7% and 66.7% of strains of the genera *Enterobacter*, *Enterococcus*, *Escherichia*, *Serratia* and *Shigella*, respectively. However, none of the *B. cereus* group strains appears to express this activity under the experimental conditions of the study carried out by [TEC 15]. Thus, while it is highly probable that *B. cereus* group bacteria are the first to be involved in spoiling phospholipase activities, they appear to be less involved in the hydrolysis of triglycerides than other types of bacteria identified in spoiled desserts made of egg products, such as the genus *Enterococcus*.

2.4.2.5. *Regulation of the expression of hydrolytic activities*

2.4.2.5.1. Bacterial factors

In the *B. cereus* group, most virulence factors – including hydrolases – secreted at the onset of the stationary phase are directly or indirectly regulated by the transcription factor phospholipase C regulator (PlcR) [GOH 05, AGA 99, SLA 02]. PlcR activates the expression of target genes by binding to a nucleotide sequence called the "PlcR box". This sequence is located upstream of the regulated genes and at variable distances from the -235 box of the sigma A promoter [AGA 99]. The transcription of PlcR occurs by auto-induction and begins just before the cells enter the stationary growth phase. It reaches a plateau after about two hours [LER 96]. The sporulation factor Spo0A represses its transcription and the transcription factor PapR is necessary for its activity. PapR is expressed under the control of PlcR, in the form of a pro-peptide. It is activated after secretion in the extracellular medium and it is in this active form that it re-enters the cell by means of an oligopeptide permease (OppABCDF) [SLA 02, GOM 01]. Thus, the three partners PlcR, OppABCDF and PapR function as a quorum sensing system [DEC 07]. PlcR is thus able to integrate two types of signals, the physiological growth stage, through Spo0A, and the cell density, through PapR [SLA 02, LER 00]. [GIL 07] measured the concentration of 22 extracellular *B. cereus* proteins during the transition phase between the exponential phase and the stationary phase. The production of 14 of these proteins (CerO, CytK, HblB, HblL2, NheA,

NheB, InhA2, NprP2, ColA, NprB, SFP, PC-PLC, PI-PLC and SMase), probably secreted via the SecA translocation machinery, is under the control of PlcR. The same team highlighted a list of 45 genes controlled by 28 PlcR boxes [GOH 08]. Forty of the PlcR-controlled proteins are exported, 22 of which are secreted into the extracellular medium and 18 bound or attached to cell wall structures (membrane or peptidoglycan). The functions of these proteins are related to the supply of nutrients (phospholipases, proteases, toxins), cellular protection (bacteriocins, toxins, transporters, cell wall biogenesis) and the perception of the environment (two-component sensors, chemotaxis proteins, GGDEF protein family regulators involved in the metabolism of a secondary bacterial messenger, cyclic di-guanosine monophosphate (di-GMPc)). Four genes code for cytoplasmic regulators. The PlcR regulon therefore appears to be a key element in the adaptation of *B. cereus* to its environment. Indeed, it integrates a wide range of environmental signals, including nutritional deficiency, self-regulation of cell density, and regulates the transcription of genes to overcome the barriers of entry into the host: nutrient source, host defenses, immune defenses and competition with other bacterial species. Another regulator of the RNPP family (Rap/NprR/PlcR/PrgX) [ROC 10], [PER 16] is the neutral protease regulator (NPrR) whose activator peptide is NprX [PER 11]. The NPrR/NPrX complex activates the expression of the *nprA* gene coding for a metalloprotease as well as other genes coding for other hydrolytic enzymes (other proteases, esterases, lipases, chitinases). The PlcR-PapR complex stimulates the transcription of the nprR-nprX operon [DUB 13].

Other bacterial factors may be involved in regulating the expression of hydrolytic activities supporting pathogenicity in the *B. cereus* group [CEU 11]: growth rate, mobility, range of temperatures for growth and toxin production (psychrotrophy or mesophily), cell type (vegetative cells or spores), and the strain, knowing that the level of production of chromosomal pathogenicity factors is more subject to variation than their presence on the chromosome [RAS 05].

2.4.2.5.2. Environmental factors

The production of toxins and the transcription of virulence factors is also dependent on environmental conditions: temperature, pH, redox potential, oxygenation conditions, and availability of nutrients (nature of the carbon source, concentration of the carbon source, availability of amino acids, iron, and fatty acids). Figure 2.5. shows a potential and not yet fully described regulatory network of a number of virulence factors described for the *B. cereus* group, including proteases and phospholipases potentially involved in food spoilage. However, the significantly high diversity of the group in terms of virulence is still unexplained, despite the amount of work devoted to the study of the genome variability of bacteria in the *B. cereus* group [CEU 11].

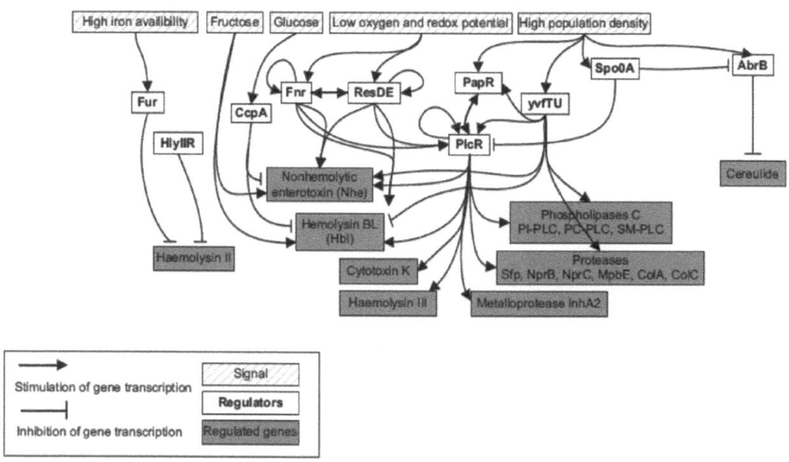

Figure 2.5. *Diagram showing the potential not yet fully described regulatory network of a number of virulence factors described for the* B. cereus *group [CEU 11]*

According to [CAR 13], some strains have the ability to grow under acidic conditions (pH 4.3) and low water availability (a_w 0.929). In many adverse conditions (acidity, high temperature, low a_w, presence of molecules with antimicrobial activity), bacteria of the *B. cereus* group are able to induce defense mechanisms against oxidative stress such as expression of catalase and superoxide dismutase activities [MOL 11]. The temperature growth range is wide within the group, with growth being observed from 4°C [VON 99] up to 50°C [GUI 08]. Despite being facultative anaerobes, this bacterial group shows lower growth rates and lower steady-state maximum concentrations under anaerobic conditions than in the presence of oxygen [DES 12]. According to these authors, the combination of a low temperature (< 15°C) and an absence of oxygen inhibits growth, probably due to the inability of the bacteria, under anaerobic conditions, to synthesize the unsaturated fatty acids essential for the adaptation of the fluidity of their membrane to low temperatures. Desaturases, however, would increase the number of unsaturations in the fatty acids of the membrane, allowing it to maintain its fluidity at low temperatures [DES 12, BRI 10, CIF 13]. Unsaturated fatty acids were detected in the bacterial membrane of *B. cereus* group bacteria incubated in a growth medium supplemented with lecithin [DES 13], increasing membrane fluidity and enhancing growth at low temperatures and under anaerobic conditions. Other mechanisms play a role in increasing the fluidity of the membrane at chilled temperatures, one of which involves a lipase whose activity modifies the proportion of branched fatty acids in the membrane fatty acids [BRI 10]. RNA helicases also play a fundamental role in adapting *B. cereus* group bacteria to cold conditions [BRO 10, PAN 10, PAN 11]. In general, the environmental conditions significantly impact the nature of the fatty acids composing the membranes of *B. cereus* group bacteria. This strong ability to modify the membrane fatty acid profiles might contribute to the ubiquity of this bacterial group [DIO 15]. However, while the modification of the membrane fatty acid composition is one of the first adaptive responses to environmental fluctuations [DES 12, SIN 71, YAN 98], certain "signatures" make it possible to differentiate the strains according to their growth temperature domains/their ecological niche [DIO 15].

2.4.3. *Metabolism of enterococci*

2.4.3.1. *Central metabolism*

Enterococci are able to metabolize 15 to 30 different sugars depending on the species. They all share the ability to use lactose, glucose, fructose, galactose, maltose, mannose, N-acetylglucosamine, amygdalin, arbutin, cellobiose, β-gentiobiose, β-D-methyl glucopyranoside, ribose, trehalose and salicin [SCH 84]. Unlike streptococci, they are not confined to the use of sugars as carbon sources. They can in fact use glycerol, lactate, malate, citrate, branched-chain α-ketoacids (α-ketoisocaproate, α-ketoisovalerate, α-keto-β-methylvalerate) and amino acids such as arginine as energy sources. Enterococci are homofermentative. The final products of their metabolism vary according to the oxygenation conditions: lactic acid is the main compound in anaerobiosis whereas it is acetate and CO_2 under aerobic conditions [AGU 12].

Under aerobic conditions, *E. faecalis* is able to achieve oxidative phosphorylation despite its inability to synthesize cytochromes unless the medium is supplemented with hemin. This species can catabolize glucose via the Embden-Meyerhof, Doudoroff or pentose phosphate pathways. Under anaerobic conditions, the main route is via the Embden-Meyerhof pathway and the main fermentation product is lactate. This is the same for *E. faecium*, but for this species, the fermentation also leads to the production of acetate, formate and ethanol in a greater quantity than the trace amounts produced by *E. faecalis*. *E. faecalis* differs from *E. faecium* by the production of acid from glycerol, the fermentation of pyruvate, the use of menaquinones as non-cytochromic electron donors, and the fact that it has no auxotrophy for folic acid [LEB 06].

The fermentation of pyruvate can take place under aerobic or anaerobic conditions. The fermentation of glycerol takes place under aerobic or microaerophilic conditions. In the genus *Enterococcus*, only the species *E. faecalis* can use glycerol as an energy source under anaerobic conditions. The use of lactose by *E. faecalis* and *E. faecium* is observed only under anaerobiosis conditions. The simultaneous presence of glucose and lactate makes it possible to increase the growth yields. With regard to the catabolism of amino acids, only *E. faecalis* is able to use arginine, agmatine and L-serine as energy sources for growth. However, the degradation of arginine, agmatine, tyrosine and phenylalanine is possible for all, half and the majority of enterococci, respectively [LEB 06].

Literature studies show the ability of enterococci to catabolize citrate [MOR 06]. Since citrate is a strongly oxidized substrate, no reducing equivalent, such as NADH, is produced during its catabolism, which results in the formation of end products other than lactic acid: diacetyl, acetaldehyde, acetoin and 2,3-butanediol, which are known to contribute to the flavoring of cheeses [HUG 93]. The catabolism of citrate is also accompanied by the production of CO_2 which contributes to the textural qualities of certain dairy products. [URD 95] show that strains of *E. faecalis* are able to produce acetate from citrate. The work of [FRE 99] and [SAR 01b] show that in the presence of sheep or goat milk, strains of *E. faecalis* and *E. faecium* are able to produce large amounts of lactic acid along with small amounts of formate and acetate. These results indicate that lactose and citrate could be co-metabolized. However, the presence of glucose prevents the consumption of citrate by a catabolic repression phenomenon, which is not the case when sucrose or galactose are present in the medium [REA 03a, REA 03b].

2.4.3.2. *Hydrolytic activities of enterococci*

Enterococci are known to produce three types of enzymes that contribute to their opportunistic pathogenicity: hyaluronidase, gelatinase and serine protease [JET 94, SEM 03, FIS 09]. The hyaluronidase, encoded by the *hyl* chromosomal gene, degrades the main constituent of the extracellular matrix of host cells, hyaluronic acid, facilitating tissue invasion and the spread of toxins [KAY 04].

Gelatinase, encoded by the chromosomal *gelE* gene, is a zinc metalloprotease active on gelatin but also on collagen, casein, hemoglobin, β-insulin and other biologically active peptides [FIS 09, KOC 04]. This enzyme also contributes to the process of biofilm formation, which may facilitate tissue colonization and persistence at sites of infection [DEL 07], [AND 16]. The serine protease, encoded by the gene *sprE*, located upstream of *gelE* and co-transcribed with the latter, also participates in pathogenesis by degrading host tissues [FIS 09].

Enterococci contribute to the development of cheese flavors during ripening due to their proteolytic and lipolytic activities and their ability to produce volatile or long-chain fatty acids, diacetyl, acetoin and other volatile compounds [FRA 99, JEN 75, SAR 01, JEN 75, ORD 78, TRO 87, CEN 96, CEN 99, CAS 97, FRA 03, SAR 01, SAR 02a]. According to [SAR 02a], *E. faecium* is involved in the development of taste, aroma, color and texture. It therefore contributes to the overall sensory profile of certain cheeses. These qualities also make it a flora of choice for other fermented products such as vegetables or meats [GIR 02, FRA 99].

While the proteolytic activities of lactic acid bacteria are well described in the literature on fermented products, those of enterococci are less well-known [FRA 03, SUZ 00, CAR 92, SOM 69, WAL 70, HEG 90, DE 91b, MOR 06]. They are rather low in dairy products, except for some strains of *E. faecalis* [SAR 01, DOV 70 ARI 97, MAC 97, SUZ 00, AND 01]. According to [CAR 92], within this genus, the level of proteolytic activity depends on the species and on the strain inside the same species. The best described proteases are able to hydrolyze casein but also β-lactoglobulin and α-lactalbumin [SAR 01, AND 01, SOM 69, WAL 70, HEG 90, DE 91a, DE 91b, VIL 94]. Many studies [JEN 95, CEN 99, CAS 97, GAR 97, OUM 01] show that cheeses produced from milks inoculated with *E. faecalis* have higher proteolytic activities than cheeses produced without these enzymes. Work by [VIL 94] on the growth of 24 strains of *E. faecium* and 60 strains of *E. faecalis* in skimmed milk at 37°C, show that the proteolytic activities of *E. faecalis* are much superior than those of *E. faecium*. These results are confirmed by [AND 01] and by [SAR 01a] in dairy products and by [MIL 10] on enterococci isolated from pasteurized liquid whole egg and responsible for their spoilage.

The subsequent degradation of amino acids is considered to have a major impact on the development of the aroma of cheeses. Proteolytic activities may also be responsible for the formation of biogenic amines by enterococci from the amino acids they (or other bacterial types) have released [GIR 02, [OGI 08, BEN 16]. Figure 2.6 shows the different biogenic amines potentially formed in foods being spoiled.

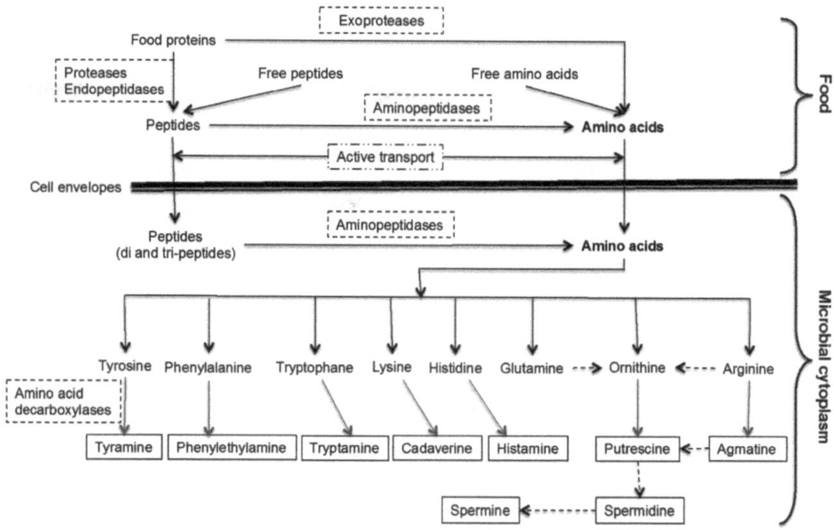

Figure 2.6. *Formation of biogenic amines in food matrices in response to microbiological metabolic activities.The gray arrows represent the decarboxylation reactions directly leading to the formation of biogenic amines, the dotted lines indicate the biogenic amines produced via metabolic pathways that differ from the one-step decarboxylation reactions. (Figure taken from [BEN 16], adapted according to [RUI 04])*

Histamine, tyramine, phenylethylamine and cadaverine are produced by a single step decarboxylation reaction from their respective precursors histidine, tyrosine, phenylalanine and lysine. The production of these biogenic amines, followed by their secretion, requires (1) active transport systems of the precursor amino acids, (2) decarboxylations and (3) amine excretion systems resulting from these decarboxylations. The transport generally involves a protein which exchanges the precursor amino acid with the amine resulting from its decarboxylation. Once in the cell, the amino acid undergoes a decarboxylation catalyzed by a specific decarboxylase (carboxy-lyases EC 4.1.1.1.) in the presence of pyridoxal phosphate. The most studied decarboxylases are histidine decarboxylase (HDC), tyrosine decarboxylase (TDC) and lysine decarboxylase

(LDC), respectively coded by the *hdcA*, *tdcA* and *cadA* genes. They later are organized into operons with other genes involved in other stages of the production process of biogenic amines, such as transport and maturation. The specificity of these decarboxylases has long been debated and it is now well established that a decarboxylase can decarboxylate several substrates. For example, *Enterococcus* TDC can decarboxylate phenylalanine and tyrosine to produce phenylethylamine and tyramine, respectively [COL 11, MAR 06, MAR 12]. Putrescine, agmatine, spermidine, and spermine are synthesized by different pathways catalyzed by groups of genes that may be specific to the species or strain because they are potentially acquired by horizontal transfer [MAR 12, LAD 11, LIN 11, WUN 14].

Enterococci are described as being capable of causing decarboxylase activities responsible for the production of biogenic amines in fermented foods. They are even recognized as having the strongest tyrosine and phenylalanine decarboxylase activities in the bacterial world [SUZ 03]. Lactic acid bacteria in the endogenous flora of fermented meats such as sausages are recognized as the major tyramine producers [ANS 02]. Cheeses [THA 90, GIR 95, GAR 01] and some meat products [BOV 01] may be suitable substrates for the production of 2-phenylethylamine, tyramine, histidine or other amines. These strains therefore harbor genes or operons encoding decarboxylases or other enzymes involved in the synthesis of biogenic amines or in their catabolism [BEN 16, WUN 14, KOM 08, NOU 94, KOU 10]. For seafood products, the synthesis of biogenic amines is not attributed to enterococci [VAL 10]. While heat-resistant bacteria such as enterococci, which are the main bacteria involved in the production of biogenic amines in dairy products [LAD 11, SPA 10], are only partially eliminated by pasteurization, heat treatment reduces the level of biogenic amines. This phenomenon may be due to the decrease in the level of contamination of the food and/or to the fact that an essential decarboxylation cofactor has been denatured by pasteurization [BEN 16].

Although lactic acid bacteria are not very active towards lipids, they are known to participate in the flavoring of cheeses, either by conversion of short chain fatty acids into ketones or methylated lactones, or by the formation of aldehydes resulting from the oxidation of unsaturated fatty acids and responsible for the development of rancid odors and tastes, either by solvation of the aromatic compounds produced from lipids but also from proteins and lactose [FOU 06]. According to [TSA 94b, TAV 98], and [HEM 98], enterococci have esterase activities that are often stronger than other lactic acid bacteria. Among the enterococci, these activities are also variable and research by [SAR 01a] show that they are stronger in *E. faecalis* than in *E. faecium*. Other studies [SUZ 00, DUR 01] show that the lipolytic activity of enterococci varies not according to the species, but to the type of cheese from which the strain originates [FOU 06].

As early as 1965, [LUN 65] showed esterase activities expressed by *E. faecalis* and *E. faecium* strains. [DOV 70] shows that *enterococci* were more active on triglycerides than *Streptococcus* strains and that the hydrolysis rates decreased with increasing the carbon chain length of the released fatty acids (tripropionin > tributyrin > tricaprin > tricaprylin). They highlight the lack of activity on triolein. [VIL 94] show that enterococci exhibit low lipolytic activity when placed in a growth medium containing whole milk. Work by [MAR 76] shows tributyrin hydrolysis expressed by *E. faecium* and *E. faecal is* isolated from cheese, yet these strains are not active on milk lipids. [CHA 79a, CHA 79b] show that a lipase of *E. faecalis*is also more active on tributyrin than on tricaproin, tricaprylin andtriolein. However, according to [CAR 92], the ability of enterococci to hydrolyze milk lipids is highly variable from one strain to another. Some strains show no lipolytic activity whereas others show high activity on a wide range of fatty acids. Work by [TSA 93], carried out on synthetic substrates (4-nitrophenyl-acyls), show *E. faecium* activity on fatty acids whose chain length is between 2 and 18 carbons. In subsequent studies [TSA 94], these same authors show that an intracellular esterase of *E. faecium* develops, *in vitro*, an activity on 4-nitrophenyl-acyls whose chain length is between 2 and 12 carbons, with an optimum in the presence of 4-nitrophenyl-butyrate (C2) [TSA 94]. Work by [SAR 01a] show that the majority of enterococci strains tested (90%, N = 129), are able to hydrolyze homogeneous triglycerides bearing fatty acids with a chain length of between 4 and 18 carbons, the activity decreasing with increasing carbon chain length. All the strains tested (100%) in this same study show an activity on the synthetic substrates, from 4-nitrophenylacetate (C2) to 4-nitrophenylstearate (C18), with the activity also being reduced by an increase in the length of the fatty acid carbon chain.

Among the few published studies on hydrolytic activities that may be involved in the spoilage of egg products are those of [MIL 10] which show lipolytic activities expressed by the strains *E. faecium* and in particular *E. faecalis* isolated from spoiled pasteurized liquid whole egg products. Work by [TEC 15] reveal that lipase activities are also detected for half of the *Enterococcus* strains derived from industrial egg whites and tested for their ability to spoil one of the main components of the floating island desserts, the custard cream.

2.4.3.3. *Regulation of hydrolytic activities of enterococci*

2.4.3.3.1. Bacterial factors

The production of cytolysin by enterococci is regulated by a quorum sensing mechanism involving a two-component system. It is the same for the production of gelatinase and serine protease. Indeed, the transcription of *gelE* and *sprE* is regulated by a system known as the Faecal Streptococci

Regulator (fsr) locus [SIF 02]. The locus comprises the genes *fsrA*, *fsrB* and *fsrC* which are co-transcribed. The gene *fsrB* encodes an enzyme that releases an autoinducing peptide (gelatinase biosynthesis activating pheromone, GBAP) whose accumulation results in a transition from the exponential phase to the stationary phase and the induction of the expression of the *gelE* and *SprE* genes.

Several teams describe the genetic basis for regulating the production of biogenic amines in enterococci [MAR 06, DRI 88, SUÁ 13, BAR 02, KAU 15, LAD 12].

2.4.3.3.2. Environmental factors

The ability of enterococci to survive many stresses [VAN 06] may explain the occurrence of cross-contaminations between food, environment and hospital environments [FIS 09].

Enterococci have growth temperature and pH ranges between 5 and 50°C and 4.6 and 9.9, respectively [FIS 09]. What differentiates *E. faecalis* and *E. faecium* from streptococci is their ability to withstand 30 min heat treatment at 60°C [FOU 06]. Their thermoresistance depends not only on the magnitude of the thermal shock but also on the physiological stage of growth at which heat stress is applied. Their optimum growth temperature is 42.7°C on an aerobic BHI agar medium with an optimum pH of 7.5. The temperature resistance can be explained by the lipid composition of their membranes. Indeed, membrane stability is greatest at the minimum growth temperature, which is a specific mechanism for enterococci [IVA 99]. At higher temperatures, the content of membrane fatty acids increases and the proportion of unsaturated fatty acids decreases.

[CAM 44] and [GUN 44] showed that pH also has a very strong effect on citrate catabolism by the genus *Enterococcus*.

E. faecalis and *E. faecium* withstand concentrations of 40% (p/v) bile salts. *E. faecalis* is able to grow in the presence of 6.5% NaCl. It exhibits cation homeostasis that contributes to its resistance to pH changes, salt and metal concentrations, and desiccation [FIS 09]. Low concentrations of NaCl (2%, p/v) enhance casein hydrolysis whereas high concentrations inhibit it in *E. faecalis* [HEG 90].

[WES 90] showed that more than half of the psychrotrophic strains of *E. faecium* and *E. faecalis* that they isolated from dairy products exhibited proteolytic activities at low temperature.

The production of biogenic amines in foods depends on many factors, including the availability of the precursors, the amino acids. Other influential factors are the availability of pyridoxal phosphate, a coenzyme involved in decarboxylation reactions [EDW 81], pH, temperature, salt concentration and a_w [GAR 16]. The transcription of genes encoding decarboxylases are induced under acidic conditions and are part of the responses of bacteria when subjected to acid stress [PER 09, ROM 12, ROM 07, PER 15].

E. faecium and *E. faecalis* can produce biogenic amines at pH 4, with the optimum being between pH 5.5 and 6 [LIU 14]. The optimal growth temperatures coincide, for some strains and some decarboxylases, with the optimal production of biogenic amines, though not systematically [GAR 16]. [MAC 06] consider that the growth of bacteria that can produce biogenic amines is an essential condition but not sufficient for this production. Moreover, decarboxylases can remain active whereas producing bacteria are no longer involved [GAR 16]. In any case, refrigeration can limit both bacterial growth and the production of biogenic amines, although it is useful to specify that psychrotrophic strains can produce them at refrigeration temperatures [EMB 06, KAN 07]. In general, the increase in salt concentration contributes to the reduction of biogenic amine synthesis by bacteria, mainly due to the reduction of decarboxylase activities [GAR 16]. Inhibitory growth concentrations, however, do not always match with the concentrations inhibiting decarboxylase activities [GAR 16].

In two consecutive publications, [CHA 79a, b] show the effect of the nature of the fatty acids present in the culture on the growth and production of a lipase (21 kDa, pI = 4.6) by *E. faecalis*. Whereas short chain fatty acids (C3, C4, C6 and C10) activate the production of this lipase, the long chain saturated (C12, C14, C16, C18) and unsaturated (C18: 2) fatty acids tested in these studies inhibit its synthesis. The maximum activity of this enzyme is observed at pH 7.5 and 40°C. [TSA 94a] show optimal activity for anintracellular esterase of *E. faecium* (45 kDa) at pH 8 and 35°C.

2.4.4. *Quorum sensing in spoilage phenomena*

Bacteria communicate with each other and with other cell types (symbionts, competitors, host cells), which allows them to coordinate their metabolic activities of survival and species differentiation. Quorum sensing (QS) is one of the mechanisms underlying these interactions. It involves the synthesis, secretion and recognition of small molecules diffusing in and out of cells. As the density of the bacterial population increases, the concentration of these signaling molecules increases in the environment and, once a critical concentration threshold is reached, the activity of a target kinase or regulator is modulated. The consequence is the modification of the expression of genes which are under the control of this system, allowing a coordinated response for the benefit of the entire population. This response converges towards the acquisition of genetic skills, virulence, sporulation, biofilm formation, hydrolytic enzyme synthesis, conjugative plasmid transfer, symbiosis and/or synthesis of antimicrobial peptides [BAS 06, DUN 97, KLF 01, MAR 04, SHA 98, WIL 07, WIL 07b, SMI 04].

Four main types of communication molecules, known as autoinducers or pheromones are described: (1) N-acyl homoserine lactones (AHLs) or autoinducers-1 (AI-1) which are fatty acid derivatives produced and used by Gram-negative bacteria, mainly for their intra-specific communication [SMI 06, PAP 16]; (2) furanosyl borate diester, or autoinducer-2 (AI-2), derived from the transformation of S-adenosyl-homocysteine into homocysteine, which is produced by Gram-positive and Gram-negative bacteria and is assumed to serve as an universal signal of intra- and inter-specific communication [DEK 06, WIN 03]; (3) autoinducer-3 (AI-3) of enterohemorrhagic *E. coli* (EHEC), which serves as a signal of virulence and interaction with the host [REA 07, SPE 03] and (4) self-inducing peptides (SIPs) of Gram-positive bacteria [LYO 04, STU 02]. SIPs are small in size (from 5 to 26 amino acids) and are highly stable, highly specific and highly diverse. They can be linear or cyclical [WIN 03]. These peptides, synthesized in an inactive form, are matured and secreted via an ATP cassette (ATP-binding cassette (ABC) transporter). There are two categories of SIPs according to whether they are recognized by a membrane sensor exposed on the surface of the cell or by an intracellular sensor [WIN 03, STO 02] (Figure 2.7). These peptides thus make it possible to trigger a direct response, by binding to a membrane protein of the recipient cell, or an indirect response after internalization, via a two-component system (Figure 2.7) [SLA 14]. The genera *Bacillus* and *Enterococcus*, which we focus on in this chapter, are described as using QS to develop genetic competence, produce antimicrobial peptides or exotoxins and to form biofilms [POD 04]. *E. faecalis* uses a two-component system similar to the *Staphylococcus* Agr system, which allows for detecting the presence of SIPs. Once the SIPs are detected, the bacteria produce and secrete two extracellular proteases, a gelatinase and a serine protease [QIN 00].

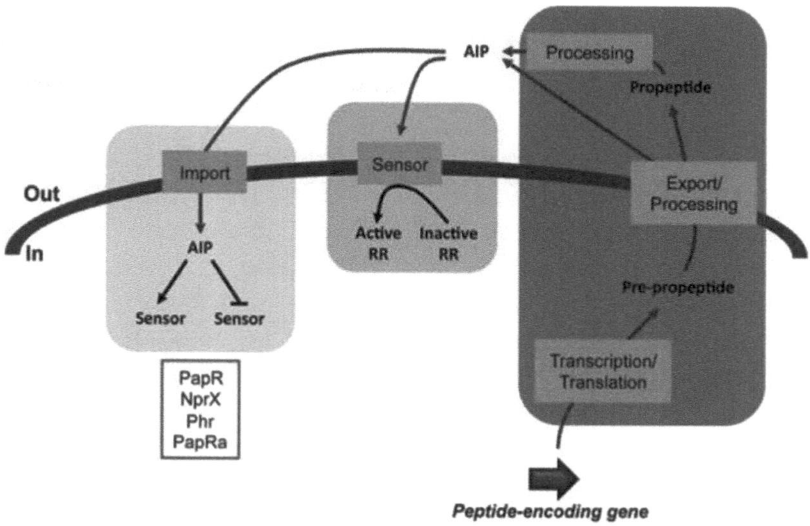

Figure 2.7. *Schematic representation of two quorum sensing systems involving SIP in Gram-positive bacteria. The step contained in the right panel is common to both systems, and the middle panel shows the dependent SIP extracellular mechanism. The left panel shows the intracellular SIP dependent mechanism. The black arrows represent positive effects, the cut black lines represent the negative effects. The purple lines indicate the circuit followed by the SIP. RR: response regulator. Figure taken from [SLA 14]. For a color version of this figure, see www.iste.co.uk/goncalves/ovoproducts.zip*

The mechanisms of direct interaction between a signaling peptide and a receptor protein can be illustrated by the example of the Rap proteins in *B. subtilis* [PER 96, SOL 96], the transcriptional activator PlcR in *B. cereus* [AGA 99], [SLA 02], and the transcriptional repressor PrgX in *E. faecalis* [BAE 04, SHE 05]. These effectors and the NprR protein of *B. cereus* are members of the QSRNPP (for Rap, NprR, PlcR and PrgX) family of proteins [DEC 07, ROC 10]. The binding of Rap and PrgX to their respective signaling peptides, Phr and cCF10, causes the deactivation of the regulatory functions of Rap and PrgX. In contrast, the binding of the signaling peptide PapR activates the transcriptional activator PlcR. NprR is a transcriptional regulator whose activity depends on the NprX signaling peptide. By binding to NprX, NprR activates the transcription of an extracellular protease (NprA) during the sporulation process [PER 11].

While the involvement of QS in bacterial pathogenicity phenomena are well described [SMI 04], it is also becoming increasingly described in food spoilage phenomena. Bacterial hydrolytic enzymes are typically produced at the end of the exponential growth phase, when the population tends to quorum [RAJ 02, STE 04]. Many enzymes, including lipases and proteases involved in food spoilage, have been identified as being QS-regulated [LIU 03, LIU 07, RIE 01, WHI 00]. In addition, some QS-specific signaling molecules have been detected in spoiled foods [LIU 03, SWI 01, BAI 11, LIN 98]. This is the case of type AI-1 and AI-2 autoinducers found in milk [WHI 00, PIN 07, LU 04, DUN 05], meat [BRU 04, JAY 03, FLO 03, LIU 06], seafood [GRA 02, CON 10, FLO 05, CHR 03] and vegetables [LIA 89, PIR 93]. However, the involvement of SIPs in food spoilage has, to our knowledge, not been described.

Knowing that spoilage almost always involves the development of a mixed bacterial population in food, the production of QS molecules is probably a determining factor in the succession of different microbial communities. The concepts of specific spoilage organisms (SSOs) that dominate the food and ephemeral spoilage organisms (ESOs), which can become dominant in the food because they are selected during food storage, are used to describe the fact that, in most cases, only a fraction of the bacterial population temporarily dominates in a spoiled food ecosystem [INT 96, GRA 02, BOA 83, NYC 08]. Therefore, different QS systems can coexist and act in synergy or compete with one another. The molecules involved in the composition of food can both inhibit and activate QS systems. The microarchitecture of the food, and both biotic and abiotic factors, can also influence intercellular communication and hence spoilage mechanisms

[AMM 08]. The literature also describes interference strategies among competing microorganisms in heterogeneous bacterial communities. These strategies include the degradation or inactivation of QS signaling molecules, phenomena known as quorum quenching, mimicry or interception of signaling molecules, or the production of toxic signaling molecules for competitors [DUB 12].

2.5. Synthesis and demonstration of potential markers of whole egg spoilage

To close this chapter, we will present a panel of molecules described as being produced during the catabolism of sugars and amino acids of enterococci and/or *B. cereus* group bacteria that we could consider using as markers of spoilage of egg products. The studies that led to the establishment of microbiological criteria in the egg product industry based on succinic, lactic and 3-hydroxybutyric acid concentrations will then be discussed. Other markers have also been proposed, some of which have not been subject to any regulatory application, but for which methods of determination have been proposed.

2.5.1. *Potential volatile markers*

Most of the research on spoiled enzyme activities focuses on dairy products and meats. Egg products are rich in protein and lipid substrates that can lead to the formation of undesirable molecules responsible for spoilages. It is possible to suspect the production of peptides or fatty acids that can impart a bitter taste including molecules derived from the catabolism of fatty acids, such as methyl ketones and lactones, or molecules derived from the oxidation of fatty acids such as highly aromatic aldehydes, as shown by the many published works on the spoilage of dairy products. This section proposes to summarize the molecules already described in the literature and potentially produced by enterococci and

B. cereus group bacteria, from sugars and amino acids. Reference is made to the spoilage of other types of food besides egg products for which literature is scarce. Table 2.4 lists the volatile compounds produced by enterococci and group *B. cereus* bacteria mentioned in the literature.

	***Enterococcus* sp.**	***B. cereus* Group**
From sugar catabolism		
3-hydroxybutanone (acetone)	Fermented meat [LAT 11]	Culture medium: [KOM 65]
	Cheese: [NIE 11, STE 09, RAN 07]	
2,3-butanedione (diacetyl)	Cheese: [NIE 11, STE 09, RAN 07]	–
Acetic acid	Fermented meat: [LAT 11]	Culture medium [ROS 05, DUP 06]
	Cheese: [STE 09]	
1-hexanol	Fermented meat: [LAT 11]	[CAS 15]
	Cheese: [RAN 07]	
Hexanoic acid	Cheese: [STE 09]	–
Butanoic acid	Cheese: [RAN 07]	–
2-butanone	Fermented meat: [LAT 11]	–
2-heptanol	Fermented meat: [LAT 11]	Fermented meat: [CAS 15]
From amino acid catabolism		
2-methylpropanol	Culture medium: [TAV 02]	–
3-methylbutanal	Fermented meat: [LAT 11]	–
	Cheese: [STE 09]	
Dimethyl disulfide	Fermented meat: [LAT 11]	Fermented meat: [HUA12] Culture medium: [HUA 12]
Benzaldehyde	Fermented meat: [LAT 11]	–
	Culture medium: [TAV 02]	
Dimethly trisulfide	Culture medium: [TAV 02]	–
Ethylpropanoate	Cheese: [TAV 02, RAN 07]	–
Ethylbutanoate	Cheese: [TAV 02, RAN 07]	–

Table 2.4. *Literature data on the volatile compounds produced by enterococci and the B. cereus group*

In spoiled meat, cheesy or creamy odors have been associated with the production of volatile compounds such as acetone [CAS 15]. Alcohols are generally derived from the catabolism of sugars and are responsible for characteristic odors such as "cut grass" for hexanol [CAS 15]. Mechanisms for the production of volatile compounds from amino acids have been particularly well described in the development of the flavoring of cheese or other fermented products [MCS 00, CAN 01]. The intracellular amino acid pool is then partially transformed into aromatic compounds through the action of several enzymes including transaminases, decarboxylases, dehydrogenases and lyases. These aroma compounds include alcohols, aldehydes, acids, esters and sulfur compounds [MCS 00]. These compounds are then excreted by passive diffusion through bacterial membranes [SMI 14]. Methylated compounds such as 2-methylpropanol and 3-methylbutanal are generally derived from leucine and isoleucine catabolism [SMI 09]. These can come from the hydrolysis of proteins, the catabolism of threonine or be synthesized from pyruvate [SMI 09]. 2-methylpropanol is associated with alcoholic odors [CAS 15]. Sulfur compounds, such as dimethyl trisulfide, from the catabolism of amino acids, are generally associated with cauliflower/onion/broccoli odors [CAS 15]. 2-methyl-propanol (isobutanol) is derived from the catabolism of amino acids and especially from the catabolism of valine. It is usually associated with alcoholic odors [MON 98]. Propanoate and ethyl butanoate are produced from propanoic and hexanoic acids, respectively, and are likely to be derived from the catabolism of sugars.

2.5.2. *Organic acids as microbiological criteria*

The European regulation establishes concentration limits for certain compounds in pasteurized egg products such as 3-OH-butyric acid, lactic acid and succinic acid. The regulatory provisions for egg products are briefly presented in Chapter 3 (section 3.2); the origin and history of these markers is discussed below:

3-hydroxybutyric acid

Eggs from incubators are sometimes discarded after up to 18 days of incubation if they are clear or not fertile. These eggs are called incubator reject eggs [HEA 76]. The use of these eggs for the manufacture of egg products was authorized until 1986 until prohibited by the European Commission Regulation No. 3494/86 (13/11/86). Various authors have proposed 3-hydroxybutyric acid as a chemical indicator of the fraudulent use of incubator reject eggs for the production of egg products [HEA 76, SAL 72, STI 87, MOS 95, CAR 02, ALA 04]. According to these authors, this compound is solely related to embryonic development and not to any microbiological activity. The concentration of 3-hydroxybyutyric acid appears to be affected by neither the incubation procedures nor by the heat treatment processes applied to egg products [SAL 72, JON 79, ROB 75]. 3-hydroxybutyric acid therefore appears to be a specific marker of embryonic development in the egg and is therefore exclusively suitable for the detection of products made from incubator reject eggs.

Succinic acid

The succinic acid present in eggs is the consequence of bacterial metabolism and its concentration evolves during egg storage [MOS 95, ALA 04, YOR 71, 72, SUM 07]. According to [YOR 71] and [YOR 72], the production of succinic acid is thought to be related to the growth of bacteria of the genera *Streptococcus* and *Salmonella* and of the species *E. coli*. According to [SUM 07], succinic acid can be considered to be a relevant marker of the microbiological quality of raw materials used for the manufacture of egg products. The use of this indicator of bacterial contamination is, however, questioned by other authors; [HID 04] show, for example, the variation in the concentration of succinic acid observed after heat treatments; [ALA 04] point out that the concentration of succinic acid is difficult to interpret because of the heat-lability of this compound.

The literature provides conflicting information on the evolution of succinic acid concentration in eggs and egg products during storage. According to [STI 87], the amount of succinic acid varies from around 8 mg to 90 mg per kg of dry matter between fresh and stale eggs. [ALA 04] showed that the amount of succinic acid was the same in fresh eggs and in eggs stored at 37°C for 18 days. Finally, [HID 04] point out that the amount of succinic acid was of the same order for Class A eggs from conventional commercial circuits as for Class B eggs kept at 4°C for six months. [ALA 04] conclude that a high concentration of succinic acid may be associated with illegal practices in the context of egg product manufacturing.

Lactic acid

Lactic acid is naturally present in eggs [HID 08, HEA 76, NEE 31]; it is also recognized as an indicator of the microbiological degradation of eggs [HID 08, MOS 95, CAR 02, YOR 71, YOR 72, SUM 07, ROS 99] and of embryonic development [STI 87, MOS 95, ROS 99, 10]. Some authors such as [SUM 07] consider this compound to be a relevant indicator of the microbiological quality of the raw materials used (analysis before pasteurization). Others, including [HID 04], question its relevance and highlight the low sensitivity of this marker. In their study, these authors show that lactic acid can only identify part of the products prepared with contaminated raw materials and they conclude that this marker should only be used for the analysis of egg products after pasteurization and not on raw materials as required by European legislation.

Lactic acid concentrations of less than 200 mg/kg of material are expected in high-quality liquid whole egg products [SUM 07]; small amounts of this compound have also been detected in pasteurized liquid samples of high microbiological quality (with bacterial levels lower than 2 log cfu/mL) [YOR 71, YOR 72]. [CAR 02] found similar concentrations of lactic acid before and after pasteurization and concluded that the microbiological load is not correlated with the concentration of this compound in the egg product matrix. [YOR 71] and [YOR 72] showed that concentrations up to 7 log cfu/mL of *E. coli* in liquid whole egg leads to the production of lactic acid at 131 mg/kg after 24 hours of incubation at room temperature. The concentration then decreased for longer incubation periods. These authors concluded that lactic acid could be produced and consumed in liquid whole egg, depending on the bacterial type involved and on the growth phase concerned. However, they did not observe a correlation between the lactic acid concentration and the level of bacterial contamination. In contrast, other authors [SUM 07, STE 68] consider lactic acid to be a relevant marker of microbiological contamination of egg products during storage; lactic acid concentrations greater than 2,000 mg/kg were detected in liquid whole egg products after 56 hours of incubation at 27°C, and the microbiological analysis of these products showed significant contamination by *Salmonella* and coliforms. According to these authors, lactic acid originates from lactic acid bacteria and from enterobacteria. [SUM 07] also found a correlation between lactic acid concentration and certain microbiological parameters (count of Enterobacteriaceae, total viable mesophilic flora) by observing the progressive degradation of a liquid whole egg product. These authors concluded that the lactic acid concentration provided relevant information on the microbiological quality of egg products prior to pasteurization.

2.5.3. *Other potential markers*

Uracil is a product of the catalytic activity of bacterial nucleoside phosphorylases on uridine [HID 08, HID 04, ALA 04]; this indicator has allowed the identification of a greater proportion of contaminated egg product samples than lactic acid. Uridine is present in varying amounts in eggs [ROS 95], and its concentration also increases during egg storage [HID 96, LUC 96]. As uracil is naturally absent in eggs [ROS 95], it was quickly proposed as a potential indicator for assessing the hygienic quality of pasteurized egg products [STI 87, MOR 87]. Uracil is also detectable in eggs released from incubators (as are 3-hydroxybutyric acid and lactic acid), where its presence is due to embryonic development [ROS 99]. [ROS 95] consider, however, that the detection threshold of uracil in egg products is exceeded when the level of bacterial contamination reaches 6 log cfu/mL; these high levels of contamination are rarely encountered in egg products stored at low temperatures or in those that are heat treated immediately after production.

Pyroglutamic acid [ROS 95] and furosine (ε-N-(2-furoyl-methyl-L-lysine)) [LUC 96, HID 95, 06] have also been described as relevant indicators of egg freshness. Pyroglutamic acid is, according to [VAN 71], derived from glutathione via γ-glutamyl cycle reactions. Furosine is produced by the acid hydrolysis of Amadori compounds, during the early stages of Maillard reactions [ERB 07]. [LUC 96] studied the evolution of furosine, pyroglutamic acid and uridine concentrations in egg white products over long storage periods at different temperatures; the pyroglutamic acid concentration was very low and increased linearly according to the temperature and shelf life (even for eggs stored at low temperatures). This increase may be due to the transfer of this compound from the egg yolk (where the concentration is higher) to the egg white during storage due to the weakening of the vitelline membrane. The concentration of furosine increased very slowly at low temperatures. Finally, the concentration of uridine increased exponentially during storage as a function of the storage temperature. At low temperature (5°C), the uridine concentration remained stable during the first five months and then increased rapidly. These authors [LUC 96] concluded that furosine was a relevant marker of egg freshness if stored at room temperature, with both high repeatability and low egg natural variability and that uridine could serve as a complementary indicator for eggs kept at low temperatures for long periods. Pyroglutamic acid was deemed an irrelevant marker because, even if it evolves linearly during storage, its concentration presents a significant natural variability and a low dynamic range of increase.

2.5.4. *Methods proposed in the literature for determining potential spoilage markers of egg products*

The search for spoilage markers of egg products is still partly unexplored. In the USA, the acceptability of liquid whole egg products is partially based on product odor, as assessed by a trained inspector. The detection of dimethylsulfide (DMS) has been proposed as an objective method for assessing the acceptability of liquid or frozen whole egg products for human consumption [TEC 13]. Other compounds such as uracil, lactic acid, acetic acid, succinic acid, pyroglutamic acid and 3-hydroxybutyric acid have been proposed as spoilage markers and different methods have been developed for their detection, as presented in Table 2.5.

Methods	Spoilage criteria/liquid egg product tested	Advantages	Disadvantages
Odor analysis	Slightly acidic; putrid/ liquid whole egg, white, yolk	sensitive, fast	Requires experts, time-consuming
Gas chromatography–mass spectrometry	Dimethylsufide (DMS)/liquid whole egg, white, yolk	sensitive	Significant technical constraints
High performance liquid chromatography	Uracil, lactic acid and acetic acid/liquid whole egg	fast	Poor sensitivity
Enzymatic kit	3-hydroxybutyric acid, succinic acid, lactic acid, uracil/liquid whole egg and fresh pâté	fast	Low extraction yields Poor sensitivity

Table 2.5. *List of available methodologies allowing for the detection or prediction of spoilage of liquid whole egg products. Compiled from [HID 08, 04, ALA 04, BRO 86, COR 08, POS 10, RIV 01]*

2.6. Bibliography

[AGA 94] AGATA N., MORI M., OHTA M. *et al.*, "A novel dodecadepsipeptide, cereulide, isolated from *Bacillus cereus* causes vacuole formation in HEp-2 cells", *FEMS Microbiol. Lett.*, vol. 121, no. 1, pp. 31–34, 1994.

[AGA 99] AGAISSE H., GOMINET M., ØKSTAD O.A. *et al.*, "PlcR is a pleiotropic regulator of extracellular virulence factor gene expression in Bacillus thuringiensis", *Mol. Microbiol.*, vol. 32, no. 5, pp. 1043–1053, 1999.

[AGU 12] AGUILAR-GALVEZ A., DUBOIS-DAUPHIN R., DESTAIN J. *et al.*, "Les entérocoques: avantages et inconvénients en biotechnologie (synthèse bibliographique)", *Biotechnol. Agron. Société Environ.*, vol. 16, no. 1, p. 67, 2012.

[ALA 04] ALAMPRESE C., "Hygienic quality evaluation of the egg product used as ingredient in fresh egg pasta", *Food Chem.*, vol. 87, no. 2, pp. 313–319, 2004.

[ALB 10] AL BULUSHI I.M., POOLE S.E., BARLOW R. *et al.*, "Speciation of Gram-positive bacteria in fresh and ambient-stored sub-tropical marine fish", *Int. J. Food Microbiol.*, vol. 138, no. 1, pp. 32–38, 2010.

[AMM 08] AMMOR M.S., MICHAELIDIS C., NYCHAS G-J.E., "Insights into the role of quorum sensing in food spoilage", *J. Food Prot.*, vol. 71, no. 7, pp. 1510–1525, 2008.

[AND 01] ANDRIGHETTO C. *et al.*, "Phenotypic and genetic diversity of enterococci isolated from Italian cheeses", *J. Dairy Res.*, vol. 68, no. 2, pp. 303–316, 2001.

[AND 16] ANDERSON A.C. *et al.*, "Enterococcus faecalis from food, clinical specimens, and oral sites: prevalence of virulence factors in association with biofilm formation", *Front. Microbiol.*, vol. 6, p. 1534, 2016.

[ANS 02] ANSORENA D. *et al.*, "Analysis of biogenic amines in northern and southern European sausages and role of flora in amine production", *Meat Sci.*, vol. 61, no. 2, pp. 141–147, 2002.

[ARI 97] ARIZCUN C., BARCINA Y., TORRE P., "Identification and characterization of proteolytic activity of Enterococcus spp. isolated from milk and Roncal and Idiazabal cheese", *Int. J. Food Microbiol.*, vol. 38, no. 1, pp. 17–24, 1997.

[ARI 12] ARIAS C.A., MURRAY B.E., "The rise of the Enterococcus: beyond vancomycin resistance", *Nat. Rev. Microbiol.*, vol. 10, no. 4, pp. 266–278, 2012.

[ASK 86] ASKAR A., TREPTOW H., *Biogene Amine in Lebensmitteln: Vorkommen, Bedeutung und Bestimmung*, Ulmer, 1986.

[AYM 06] AYMERICH T., MARTIN B., GARRIGA M. *et al.*, "Safety properties and molecular strain typing of lactic acid bacteria from slightly fermented sausages", *J. Appl. Microbiol.*, vol. 100, no. 1, pp. 40–49, 2006.

[BAC 99] BACH H.-J. *et al.*, "Specific detection of the gene for the extracellular neutral protease of Bacillus cereus by PCR and blot hybridization", *Appl. Environ. Microbiol.*, vol. 65, no. 7, pp. 3226–3228, 1999.

[BAC 01] BACH H.-J., HARTMANN A., SCHLOTER M. *et al.*, "PCR primers and functional probes for amplification and detection of bacterial genes for extracellular peptidases in single strains and in soil", *J. Microbiol. Methods*, vol. 44, no. 2, pp. 173–182, 2001.

[BAE 04] BAE T., KOZLOWICZ B.K., DUNNY G.M., "Characterization of cis-acting prgQ mutants: evidence for two distinct repression mechanisms by Qa RNA and PrgX protein in pheromone-inducible enterococcal plasmid pCF10", *Mol. Microbiol.*, vol. 51, no. 1, pp. 271–281, 2004.

[BAI 05] BAIXAS-NOGUERAS S., BOVER-CID S., VECIANA-NOGUÉS M.T. *et al.*, "Biogenic Amine Index for Freshness Evaluation in Iced Mediterranean Hake (Merluccius merluccius)", *J. Food Prot.*, vol. 68, no. 11, pp. 2433–2438, 2005.

[BAI 11] BAI A.J., RAI V.R., "Bacterial quorum sensing and food industry", *Compr. Rev. Food Sci. Food Saf.*, vol. 10, no. 3, pp. 183–193, 2011.

[BAL 16] BALAKRISHNARAJA R. *et al.*, "Studies on lipase enzyme production by indigenously isolated Bacillus Cereus (BAIT GCT 127002) and Bacillus Cereus (BAIT GCT 127001)", *J. Adv. Chem.*, vol. 12, no. 17, pp. 5010–5019, 2016.

[BAR 02] BARCELONA-ANDRÉS B., MARINA A., RUBIO V., "Gene structure, organization, expression, and potential regulatory mechanisms of arginine catabolism in Enterococcus faecalis", *J. Bacteriol.*, vol. 184, no. 22, pp. 6289–6300, 2002.

[BAR 07] BARON F. *et al.*, "Isolation and characterization of a psychrotolerant toxin producer, Bacillus weihenstephanensis, in liquid egg products", *J. Food Prot.*, vol. 70, no. 12, pp. 2782–2791, 2007.

[BAR 10] BARBOSA J., GIBBS P.A., TEIXEIRA P., "Virulence factors among enterococci isolated from traditional fermented meat products produced in the North of Portugal", *Food Control*, vol. 21, no. 5, pp. 651–656, 2010.

[BAR 14] BARON F. *et al.*, "Ovotransferrin plays a major role in the strong bactericidal effect of egg white against the Bacillus cereus group", *J. Food Prot.*, vol. 77, no. 6, pp. 955–962, 2014.

[BAR 16] BARON B. *et al.*, "Egg white versus Salmonella Enteritidis! A harsh medium meets a resilient pathogen", *Food Microbiol.*, vol. 53, pp. 82–93, 2016.

[BAS 06] BASSLER B.L., LOSICK R., "Bacterially speaking", *Cell*, vol. 125, no. 2, pp. 237–246, 2006.

[BEE 00] BEECHER D.J., WONG A.C.L., "Cooperative, synergistic and antagonistic haemolytic interactions between haemolysin BL, phosphatidylcholine phospholipase C and sphingomyelinase from Bacillus cereus", *Microbiology*, vol. 146, no. 12, pp. 3033–3039, 2000.

[BEL 80] BELLOMO G., MANGIAGLE A., NICASTRO L. *et al.*, "A Controlled Double-Blind-Study of SF68 strain as a New Biological Preparation for the Treatment of Diarrhea in Pediatrics", *Curr. Ther. Res.-Clin. Exp.*, vol. 28, no. 6, pp. 927–936, 1980.

[BEL 84] BELL R.R., LACY K.M.D., "Heat injury and recovery of Streptococcus faecium associated with the souring of chub-packed luncheon meat", *J. Appl. Microbiol.*, vol. 57, no. 2, pp. 229–236, 1984.

[BEN 04a] BEN OMAR N.B. *et al.*, "Functional and safety aspects of enterococci isolated from different Spanish foods", *Syst. Appl. Microbiol.*, vol. 27, no. 1, pp. 118–130, 2004.

[BEN 04b] BENNER JR R.A., STARUSZKIEWICZ W.F., OTWELL W.S., "Putrescine, cadaverine, and indole production by bacteria isolated from wild and aquacultured penaeid shrimp stored at 0, 12, 24, and 36° C", *J. Food Prot.*, vol. 67, no. 1, pp. 124–133, 2004.

[BEN 16] BENKERROUM N., "Biogenic amines in dairy products: origin, incidence, and control means", *Compr. Rev. Food Sci. Food Saf.*, vol. 15, no. 4, pp. 801–826, 2016.

[BOA 65] BOARD R.G., AYRES J., "Influence of temperature on bacterial infection of the hen's egg", *Appl. Microbiol.*, vol. 13, no. 3, pp. 358–364, 1965.

[BOA 77] BOARD R.G., "The microbiology of eggs", in STADELMAN W.J., COTTERILL O. (eds), *Egg Science and Technology, 2nd Edition*, AVI Publishing Co., Inc., Westport, 1977.

[BOA 83] BOARD R.G. *et al.*, *A Modern Introduction to Food Microbiology,* Blackwell Scientific Publications, Oxford, 1983.

[BOA 94] BOARD R.G., FULLER R., *Microbiology of the Avian Egg*, Springer, Heidelberg, 1994.

[BOA 00] BOARD R.G., "Eggs and egg products", in LUND B., BAIRD-PARKER T.C., GOULD G.W. (eds), *Microbiological Safety and Quality of Food*, Aspen Publisher, Maryland, 2000.

[BOR 16] BORTOLAIA V., ESPINOSA-GONGORA C., GUARDABASSI L., "Human health risks associated with antimicrobial-resistant enterococci and Staphylococcus aureus on poultry meat", *Clin. Microbiol. Infect.*, vol. 22, no. 2, pp. 130–140, February 2016.

[BOV 01] BOVER-CID S., HUGAS M., IZQUIERDO-PULIDO M. *et al.*, "Amino acid-decarboxylase activity of bacteria isolated from fermented pork sausages", *Int. J. Food Microbiol.*, vol. 66, no. 3, pp. 185–189, 2001.

[BRI 10] BRILLARD J. *et al.*, "Identification of Bacillus cereus genes specifically expressed during growth at low temperatures", *Appl. Environ. Microbiol.*, vol. 76, no. 8, pp. 2562–2573, 2010.

[BRO 84] BROCKMAN H., "Reaction scheme, interfacial structure and experimental approaches", *Lipase*, pp. 1–44, 1984.

[BRO 86] BROWN M.L., HOLBROOK D.M., HOERNING E.F. *et al.*, "Volatile indicators of deterioration in liquid egg products", *Poult. Sci.*, vol. 65, no. 10, pp. 1925–1933, 1986.

[BRO 10] BROUSSOLLE V. *et al.*, "Insertional mutagenesis reveals genes involved in Bacillus cereus ATCC 14579 growth at low temperature", *FEMS Microbiol. Lett.*, vol. 306, no. 2, pp. 177–183, 2010.

[BRU 04] BRUHN J.B. *et al.*, "Presence of acylated homoserine lactones (AHLs) and AHL-producing bacteria in meat and potential role of AHL in spoilage of meat", *Appl. Environ. Microbiol.*, vol. 70, no. 7, pp. 4293–4302, 2004.

[BUR 07] BURDYCHOVA R. KOMPRDA T., "Biogenic amine-forming microbial communities in cheese", *FEMS Microbiol. Lett.*, vol. 276, no. 2, pp. 149–155, November 2007.

[BUY 96] BUYDENS P., DEBEUCKELAERE S., "Efficacy of SF 68 in the treatment of acute diarrhea a placebo-controlled trial", *Scand. J. Gastroenterol.*, vol. 31, no. 9, pp. 887–891, 1996.

[CAD 10] CADOT C. *et al.*, "InhA1, NprA, and HlyII as Candidates for Markers To Differentiate Pathogenic from Nonpathogenic Bacillus cereus Strains", *J. Clin. Microbiol.*, vol. 48, no. 4, pp. 1358–1365, April 2010.

[CAD 12] CADEL SIX S. *et al.*, "Toxi-infections alimentaires collectives à Bacillus cereus: bilan de la caractérisation des souches de 2006 à 2010", *Bull. Épidémiologique Santé Anim. Aliment. Risques Aliment. Microbiol.*, vol. 50, pp. 57–61, 2012.

[CAM 44] CAMPBELL J., GUNSALUS I., "Citric acid fermentation by streptococci and lactobacilli", *J. Bacteriol.*, vol. 48, no. 1, pp. 71, 1944.

[CAM 06] CAMPOS C.A., RODRÍGUEZÓ, CALO-MATA P. *et al.*, "Preliminary characterization of bacteriocins from Lactococcus lactis, Enterococcus faecium and Enterococcus mundtii strains isolated from turbot (Psetta maxima)", *Food Res. Int.*, vol. 39, no. 3, pp. 356–364, 2006.

[CAN 01] CANDOĞAN K., ACTON J., "Proteolysis in sausage fermentation", *Gıda*, vol. 26, no. 4, pp. 247–253, 2001.

[CAR 92] CARRASCO DE MENDOZA M., SCARINCI H., DEL HUERTO GARAT M. *et al.*, "Technological properties of enterococci in lactic starters: acidfying and lipolytic activities", *MAN Microbiol. Aliments Nutr.*, vol. 10, no. 3, pp. 289–293, 1992.

[CAR 02] CARAZZOLO A., BALDASSINI O., BORTOLOMEAZZI R. *et al.*, "Applications of the NIRS technique on quality evaluation of eggs and egg products", *Ind. Aliment. Italy*, 2002.

[CAR 13] CARLIN F., ALBAGNAC C., RIDA A. *et al.*, "Variation of cardinal growth parameters and growth limits according to phylogenetic affiliation in the Bacillus cereus Group. Consequences for risk assessment", *Food Microbiol.*, vol. 33, no. 1, pp. 69–76, 2013.

[CAS 97] CASALTA E., ZENNARO R., "Effect of specific starters on microbiological, biochemical and sensory characteristics of Venaco, a Corsican soft cheese", *Sci. Aliments Fr.*, 1997.

[CAS 15] CASABURI A., PIOMBINO P., NYCHAS G.-J. *et al.*, "Bacterial populations and the volatilome associated to meat spoilage", *Food Microbiol.*, vol. 45, pp. 83–102, 2015.

[CEN 96] CENTENO J., MENÉNDEZ S., RODRIGUEZ-OTERO J., "Main microbial flora present as natural starters in Cebreiro raw cow's-milk cheese (Northwest Spain)", *Int. J. Food Microbiol.*, vol. 33, nos 2–3, pp. 307–313, 1996.

[CEN 99] CENTENO J., MENENDEZ S., HERMIDA M. *et al.*, "Effects of the addition of Enterococcus faecalis in Cebreiro cheese manufacture", *Int. J. Food Microbiol.*, vol. 48, no. 2, pp. 97–111, 1999.

[CEU 11] CEUPPENS S. *et al.*, "Regulation of toxin production by Bacillus cereus and its food safety implications", *Crit. Rev. Microbiol.*, vol. 37, no. 3, pp. 188–213, 2011.

[CHA 79a] CHANDER H., RANGANATHAN B., "Role of some fatty acids on the growth and lipase production by Streptococcus faecalis", *J. Food Sci.*, vol. 44, no. 5, pp. 1566–1567, 1979.

[CHA 79b] CHANDER H., RANGANATHAN B., SINGH J., "Purification and some properties of lipase from Streptococcus faecalis", *J. Food Sci.*, vol. 44, no. 6, pp. 1747–1751, 1979.

[CHE 03] CHEN L., DANIEL R.M., COOLBEAR T., "Detection and impact of protease and lipase activities in milk and milk powders", *Int. Dairy J.*, vol. 13, no. 4, pp. 255–275, 2003.

[CHE 04] CHEN L., COOLBEAR T., DANIEL R., "Characteristics of proteinases and lipases produced by seven Bacillus sp. isolated from milk powder production lines", *Int. Dairy J.*, vol. 14, no. 6, pp. 495–504, 2004.

[CHE 07] CHEN S., LILI Q., BINGZHAO S., "Purification and properties of enantioselective lipase from a newly isolated Bacillus cereus C71", *Process Biochem.*, vol. 42, no. 6, pp. 988–994, 2007.

[CHI 06] CHITLARU T., GAT O., GOZLAN Y. *et al.*, "Differential Proteomic Analysis of the Bacillus anthracis Secretome: Distinct Plasmid and Chromosome CO2-Dependent Cross Talk Mechanisms Modulate Extracellular Proteolytic Activities", *J. Bacteriol.*, vol. 188, no. 10, pp. 3551–3571, 2006.

[CHO 93] CHOW J.W. *et al.*, "Plasmid-associated hemolysin and aggregation substance production contribute to virulence in experimental enterococcal endocarditis", *Antimicrob. Agents Chemother.*, vol. 37, no. 11, pp. 2474–2477, 1993.

[CHR 03] CHRISTENSEN A.B. *et al.*, "Quorum-sensing-directed protein expression in Serratia proteamaculans B5a", *Microbiology*, vol. 149, no. 2, pp. 471–483, 2003.

[CHU 06] CHUNG M-C. *et al.*, "Secreted Neutral Metalloproteases of *Bacillus anthracis*as Candidate Pathogenic Factors", *J. Biol. Chem.*, vol. 281, no. 42, pp. 31408–31418, 2006.

[CIF 13] CIFRÉ L.C., ALEMANY M., DE MENDOZA D. *et al.*, "Exploring the biosynthesis of unsaturated fatty acids in Bacillus cereus ATCC 14579 and functional characterization of novel acyl-lipid desaturases", *Appl. Environ. Microbiol.*, vol. 79, no. 20, pp. 6271–6279, 2013.

[CLA 07] CLAVEL T., CARLIN F., DARGAIGNARATZ C. *et al.*, "Effects of porcine bile on survival of Bacillus cereus vegetative cells and Haemolysin BL enterotoxin production in reconstituted human small intestine media", *J. Appl. Microbiol.*, vol. 103, no. 5, pp. 1568–1575, 2007.

[COG 77] COGAN T., "A review of heat resistant lipases and proteinases and the quality of dairy products", *Ir. J. Food Sci. Technol.*, pp. 95–105, 1977.

[COL 11] COLLINS J.L. *et al.*, "Scientific opinion on risk based control of biogenic amine formation in fermented foods", *EFSA J*, vol. 9, p. 2393, 2011.

[CON 10] CONNELL J.L., WESSEL A.K., M PARSEK M.R. *et al.*, "Probing prokaryotic social behaviors with bacterial 'lobster traps'", *MBio*, vol. 1, no. 4, pp. e00202–e00210, 2010.

[COR 08] CORREA E., DIAZ-BARCOS V., FUENTES-PILA J. *et al.*, "Modeling ovoproduct spoilage with red LED light", *IV International Symposium on Applications of Modelling as an Innovative Technology in the Agri-Food-Chain: Model-IT 802*, pp. 265–272, 2008.

[COU 82] COUSIN M.A., "Presence and Activity of Psychrotrophic Microorganisms in Milk and Dairy Products: A Review", *J. Food Prot.*, vol. 45, no. 2, pp. 172–207, 1982.

[CUN 77] CUNNINGHAM F.E., "Egg product pasteurization", in STADELMAN W.J., COTTERILL O.J. (eds), *Egg Science and Technology*, AVI Publ Co., Westport, 1977.

[DAI 96] DAINTY R., "Chemical/biochemical detection of spoilage", *Int. J. Food Microbiol.*, vol. 33, no. 1, pp. 19–33, 1996.

[DAL 93] DALGAARD P., GRAM L., HUSS H.H., "Spoilage and shelf-life of cod fillets packed in vacuum or modified atmospheres", *Int. J. Food Microbiol.*, vol. 19, no. 4, pp. 283–294, 1993.

[DAR 94] DARTOIS V., COPPÉE J.Y., COLSON C. *et al.*, "Genetic analysis and overexpression of lipolytic activity in Bacillus subtilis", *Appl. Environ. Microbiol.*, vol. 60, no. 5, pp. 1670–1673, May 1994.

[DEC 07] DECLERCK N. *et al.*, "Structure of PlcR: Insights into virulence regulation and evolution of quorum sensing in Gram-positive bacteria", *Proc. Natl. Acad. Sci.*, vol. 104, no. 47, pp. 18490–18495, 2007.

[DEC 17] DECIMO M., BRASCA M., ORDÓÑEZ J.A. *et al.*, "Fatty acids released from cream by psychrotrophs isolated from bovine raw milk", *Int. J. Dairy Technol.*, vol. 70, no. 3, pp. 339–344, 2017.

[DEE 06] DEETH H.C., "Lipoprotein lipase and lipolysis in milk", *Int. Dairy J.*, vol. 16, no. 6, pp. 555–562, 2006.

[DEE 11] DEETH H., "Lipolysis and hydrolytic rancidity", in FUQUAY J.W., FOX P.F., MCSWEENEY P.L.H. (eds), *Encyclopedia of Dairy Sciences, 2nd Edn, Vol. 3*, Academic Press, Oxford, 2011.

[DEF 91] DE FERNANDO G.G., HERNANDEZ P., BURGOS J. *et al.*, "Extracellular proteinase fromEnterococcus faecalis subsp. liquefaciens", *Folia Microbiol. (Praha)*, vol. 36, no. 5, pp. 429–436, 1991.

[DEI 63] DEIBEL R., SILLIKER J., "Food-poisoning potential of the enterococci", *J. Bacteriol.*, vol. 85, no. 4, pp. 827–832, 1963.

[DEK 05] DE KWAADSTENIET M., TODOROV S.D., KNOETZE H. *et al.*, "Characterization of a 3944 Da bacteriocin, produced by Enterococcus mundtii ST15, with activity against Gram-positive and Gram-negative bacteria", *Int. J. Food Microbiol.*, vol. 105, no. 3, pp. 433–444, December 2005.

[DEK 06] DE KEERSMAECKER S.C., SONCK K., VANDERLEYDEN J., "Let LuxS speak up in AI-2 signaling", *TRENDS Microbiol.*, vol. 14, no. 3, pp. 114–119, 2006.

[DEK 13] DE KRAKER M.E.A., JARLIER V., MONEN J.C.M. *et al.*, "The changing epidemiology of bacteraemias in Europe: trends from the European Antimicrobial Resistance Surveillance System", *Clin. Microbiol. Infect.*, vol. 19, no. 9, pp. 860–868, September 2013.

[DEL 07] DEL PAPA M.F., HANCOCK L.E., THOMAS V.C. *et al.*, "Full activation of Enterococcus faecalis gelatinase by a C-terminal proteolytic cleavage", *J. Bacteriol.*, vol. 189, no. 24, pp. 8835–8843, 2007.

[DEM 17] DEMIRGÜL F., TUNCER Y., "Detection of Antibiotic Resistance and Resistance Genes in Enterococci Isolated from Sucuk, a Traditional Turkish Dry-Fermented Sausage", *Korean J. Food Sci. Anim. Resour.*, vol. 37, no. 5, p. 670, 2017.

[DES 12] DE SARRAU B., CLAVEL T., CLERTÉ C. *et al.*, "Influence of anaerobiosis and low temperature on Bacillus cereus growth, metabolism, and membrane properties", *Appl. Environ. Microbiol.*, vol. 78, no. 6, pp. 1715–1723, 2012.

[DES 13] DE SARRAU B. *et al.*, "Unsaturated fatty acids from food and in the growth medium improve growth of Bacillus cereus under cold and anaerobic conditions", *Food Microbiol.*, vol. 36, no. 2, pp. 113–122, 2013.

[DIO 15] DIOMANDÉ S.E., NGUYEN-THE C., GUINEBRETIÈRE M.-H. *et al.*, "Role of fatty acids in Bacillus environmental adaptation", *Front. Microbiol.*, vol. 6, 2015.

[DOL 17] DOLL E.V., SCHERER S., WENNING M., "Spoilage of Microfiltered and Pasteurized Extended Shelf Life Milk Is Mainly Induced by Psychrotolerant Spore-Forming Bacteria that often Originate from Recontamination", *Front. Microbiol.*, vol. 8, 2017.

[DOS 15] DOS SANTOS K.M.O. *et al.*, "Safety, beneficial and technological properties of Enterococcus faecium isolated from Brazilian cheeses", *Braz. J. Microbiol.*, vol. 46, no. 1, pp. 237–249, 2015.

[DOV 70] DOVAT A., REINBOLD G., HAMMOND E. *et al.*, "Lipolytic and proteolytic activity of enterococci and lactic group streptococci isolated from young Cheddar cheese", *J. Milk Food Technol.*, vol. 33, no. 9, pp. 365–372, 1970.

[DRI 88] DRIESSEN A., SMID E.J., KONINGS W.N., "Transport of diamines by Enterococcus faecalis is mediated by an agmatine-putrescine antiporter", *J. Bacteriol.*, vol. 170, no. 10, pp. 4522–4527, 1988.

[DRO 93] DROBNIEWSKI F.A., "Bacillus cereus and related species", *Clin. Microbiol. Rev.*, vol. 6, no. 4, pp. 324–338, 1993.

[DUB 12] DUBOIS T., Etude du système de communication cellulaire NprR-NprX au sein du groupe Bacillus cereus, PhD thesis, AgroParisTech, 2012.

[DUB 13] DUBOIS T. *et al.*, "Activity of the Bacillus thuringiensis NprR–NprX cell–cell communication system is co-ordinated to the physiological stage through a complex transcriptional regulation", *Mol. Microbiol.*, vol. 88, no. 1, pp. 48–63, 2013.

[DUN 97] DUNNY G.M., LEONARD B.A., "Cell-cell communication in Gram-positive bacteria", *Annu. Rev. Microbiol.*, vol. 51, no. 1, pp. 527–564, 1997.

[DUN 05] DUNSTALL G., ROWE M.T., WISDOM G.B. *et al.*, "Effect of quorum sensing agents on the growth kinetics of Pseudomonas spp. of raw milk origin", *J. Dairy Res.*, vol. 72, no. 3, pp. 276–280, 2005.

[DUP 04] DUPORT C., THOMASSIN S., BOUREL G. *et al.*, "Anaerobiosis and low specific growth rates enhance hemolysin BL production by Bacillus cereus F4430/73", *Arch. Microbiol.*, vol. 182, no. 1, pp. 90–95, September 2004.

[DUP 06] DUPORT C., ZIGHA A., ROSENFELD E. *et al.*, "Control of enterotoxin gene expression in Bacillus cereus F4430/73 involves the redox-sensitive ResDE signal transduction system", *J. Bacteriol.*, vol. 188, no. 18, pp. 6640–6651, 2006.

[DUR 01] DURLU-OZKAYA F., XANTHOPOULOS V., TUNAIL N. *et al.*, "Technologically important properties of lactic acid bacteria isolates from Beyaz cheese made from raw ewes' milk", *J. Appl. Microbiol.*, vol. 91, no. 5, pp. 861–870, 2001.

[EAT 01] EATON T.J., GASSON M.J., "Molecular Screening of Enterococcus Virulence Determinants and Potential for Genetic Exchange between Food and Medical Isolates", *Appl. Environ. Microbiol.*, vol. 67, no. 4, pp. 1628–1635, 2001.

[EDL 76] EDLUND T., SIDEN I., BOMAN H.G., "Evidence for two immune inhibitors from Bacillus thuringiensis interfering with the humoral defense system of saturniid pupae", *Infect. Immun.*, vol. 14, no. 4, pp. 934–941, 1976.

[EDW 81] EDWARDS S.T., SANDINE W.E., "Public Health Significance of Amines in Cheese", *J. Dairy Sci.*, vol. 64, no. 12, pp. 2431–2438, 1981.

[EFS 14a] EFSA PANEL ON BIOLOGICAL HAZARDS (BIOHAZ), "Scientific Opinion on the public health risks of table eggs due to deterioration and development of pathogens: Public health risks of table eggs due to deterioration and development of pathogens", *EFSA J.*, vol. 12, no. 7, p. 3782, 2014.

[EFS 14b] EUROPEAN FOOD SAFETY AUTHORITY AND EUROPEAN CENTRE FOR DISEASE PREVENTION AND CONTROL, "The European Union Summary Report on Trends and Sources of Zoonoses, Zoonotic Agents and Food-borne Outbreaks in 2012: EU summary report on zoonoses, zoonotic agents and food-borne outbreaks 2012", *EFSA J.*, vol. 12, no. 2, p. 3547, 2014.

[EHL 04] EHLING-SCHULZ M., FRICKER M., SCHERER S., "Bacillus cereus, the causative agent of an emetic type of food-borne illness", *Mol. Nutr. Food Res.*, vol. 48, no. 7, pp. 479–487, 2004.

[EMB 06] EMBORG J., DALGAARD P., "Formation of histamine and biogenic amines in cold-smoked tuna: an investigation of psychrotolerant bacteria from samples implicated in cases of histamine fish poisoning", *J. Food Prot.*, vol. 69, no. 4, pp. 897–906, 2006.

[ERB 07] ERBERSDOBLER H.F., SOMOZA V., "Forty years of furosine – Forty years of using Maillard reaction products as indicators of the nutritional quality of foods", *Mol. Nutr. Food Res.*, vol. 51, no. 4, pp. 423–430, 2007.

[ERT 07] ERTUĞRUL S., DÖNMEZ G., TAKAÇ S., "Isolation of lipase producing Bacillus sp. from olive mill wastewater and improving its enzyme activity", *J. Hazard. Mater.*, vol. 149, no. 3, pp. 720–724, 2007.

[FER 83] FERNÁNDEZ-DÍEZ M.J., "Olives", in REED G. (ed.), *Biotechnology, vol 5. Food and Feed Production with Microorganisms*, Verlag Chemie, Deerfield Beach, 1983.

[FIS 09] FISHER K., PHILLIPS C., "The ecology, epidemiology and virulence of Enterococcus", *Microbiology*, vol. 155, no. 6, pp. 1749–1757, June 2009.

[FLI 17] FLINT S., GONZAGA Z.J., GOOD J. *et al.*, "Bacillus thermoamylovorans – A new threat to the dairy industry – A review", *Int. Dairy J.*, vol. 65, pp. 38–43, 2017.

[FLO 03] FLODGAARD L.R., CHRISTENSEN A.B., MOLIN S. *et al.*, "Influence of food preservation parameters and associated microbiota on production rate, profile and stability of acylated homoserine lactones from food-derived Enterobacteriaceae", *Int. J. Food Microbiol.*, vol. 84, no. 2, pp. 145–156, 2003.

[FLO 05] FLODGAARD L., DALGAARD P., ANDERSEN J.B. *et al.*, "Nonbioluminescent strains of Photobacterium phosphoreum produce the cell-to-cell communication signal N-(3-hydroxyoctanoyl) homoserine lactone", *Appl. Environ. Microbiol.*, vol. 71, no. 4, pp. 2113–2120, 2005.

[FOE 87] FOEGEDING P., STANLEY N., "Growth and inactivation of microorganisms isolated from ultrapasteurized egg", *J. Food Sci.*, vol. 52, no. 5, pp. 1219–1227, 1987.

[FOU 06] FOULQUIÉ MORENO M.R., SARANTINOPOULOS P., TSAKALIDOU E. *et al.*, "The role and application of enterococci in food and health", *Int. J. Food Microbiol.*, vol. 106, no. 1, pp. 1–24, 2006.

[FRA 98] FRANCIS K.P., MAYR R., VON STETTEN F. *et al.*, "Discrimination of psychrotrophic and mesophilic strains of the Bacillus cereus group by PCR targeting of major cold shock protein genes", *Appl. Environ. Microbiol.*, vol. 64, no. 9, pp. 3525–3529, 1998.

[FRA 99] FRANZ C.M., HOLZAPFEL W.H., STILES M.E., "Enterococci at the crossroads of food safety?", *Int. J. Food Microbiol.*, vol. 47, no. 1, pp. 1–24, 1999.

[FRA 03] FRANZ C.M., STILES M.E., SCHLEIFER K.H. *et al.*, "Enterococci in foods—a conundrum for food safety", *Int. J. Food Microbiol.*, vol. 88, no. 2, pp. 105–122, 2003.

[FRA 04] FRANZ C., HOLZAPFEL W.H., "The genus Enterococcus: biotechnological and safety issues", in SALMINEN S., WRIGHT A., OUWEHAND A. (eds), *Lactic Acid Bacteria: Microbiological and Functional Aspects*, Marcel Dekker, New York, 2004.

[FRA 07] FRANK J.F., "Milk and Dairy Products", in DOYLE M.P., BEUCHAT L.R., MONTWILLE T.J. (eds), *Food Microbiology: Fundamentals and Frontiers, Third Edition*, American Society for Microbiology, Washington D.C., 2007.

[FRE 99] FREITAS A.C., PINTADO A.E., PINTADO M.E. *et al.*, "Organic acids produced by lactobacilli, enterococci and yeasts isolated from Picante cheese", *Eur. Food Res. Technol.*, vol. 209, no. 6, pp. 434–438, 1999.

[FRI 08] FRICKER M., REISSBRODT R., EHLING-SCHULZ M., "Evaluation of standard and new chromogenic selective plating media for isolation and identification of Bacillus cereus", *Int. J. Food Microbiol.*, vol. 121, no. 1, pp. 27–34, 2008.

[GAR 97] GARDE S., GAYA P., MEDINA M. *et al.*, "Acceleration of flavour formation in cheese by a bacteriocin-producing adjunct lactic culture", *Biotechnol. Lett.*, vol. 19, no. 10, pp. 1011–1014, 1997.

[GAR 01] GARDINI F. *et al.*, "Effects of pH, temperature and NaCl concentration on the growth kinetics, proteolytic activity and biogenic amine production of Enterococcus faecalis", *Int. J. Food Microbiol.*, vol. 64, no. 1, pp. 105–117, 2001.

[GAR 16] GARDINI F., ÖZOGUL Y., SUZZI G. *et al.*, "Technological factors affecting biogenic amine content in foods: a review", *Front. Microbiol.*, vol. 7, 2016.

[GHA 00] GHANEM E.H., AL-SAYED H.A., SALEH K.M., "An alkalophilic thermostable lipase produced by a new isolate of Bacillus alcalophilus", *World J. Microbiol. Biotechnol.*, vol. 16, no. 5, pp. 459–464, July 2000.

[GIL 89] GILMORE M.S., CRUZ-RODZ A.L., LEIMEISTER-WÄCHTER M. *et al.*, "A Bacillus cereus cytolytic determinant, cereolysin AB, which comprises the phospholipase C and sphingomyelinase genes: nucleotide sequence and genetic linkage", *J. Bacteriol.*, vol. 171, no. 2, pp. 744–753, 1989.

[GIL 07] GILOIS N. *et al.*, "Growth-related variations in the Bacillus cereus secretome", *Proteomics*, vol. 7, no. 10, pp. 1719–1728, 2007.

[GIR 95] GIRAFFA G., PEPE G., LOCCI F. *et al.*, "Hemolytic activity, production of thermonuclease and biogenic amines by dairy enterococci", *Ital. J. Food Sci.*, vol. 7, no. 4, pp. 341–349, 1995.

[GIR 02] GIRAFFA G., "Enterococci from foods", *FEMS Microbiol. Rev.*, vol. 26, no. 2, pp. 163–171, 2002.

[GIR 03] GIRAFFA G., "Functionality of enterococci in dairy products", *Int. J. Food Microbiol.*, vol. 88, nos 2–3, pp. 215–222, 2003.

[GOH 05] GOHAR M., GILOIS N., GRAVELINE R. *et al.*, "A comparative study of Bacillus cereus, Bacillus thuringiensis and Bacillus anthracis extracellular proteomes", *Proteomics*, vol. 5, no. 14, pp. 3696–3711, 2005.

[GOH 08] GOHAR M. *et al.*, "The PlcR virulence regulon of Bacillus cereus", *PLoS One*, vol. 3, no. 7, p. e2793, 2008.

[GOM 01] GOMINET M., SLAMTI L., GILOIS N. *et al.*, "Oligopeptide permease is required for expression of the Bacillus thuringiensis plcR regulon and for virulence", *Mol. Microbiol.*, vol. 40, no. 4, pp. 963–975, 2001.

[GRA 90] GRAM L., WEDELL-NEERGAARD C., HUSS H.H., "The bacteriology of fresh and spoiling Lake Victorian Nile perch (Lates niloticus)", *Int. J. Food Microbiol.*, vol. 10, nos 3–4, pp. 303–316, 1990.

[GRA 93] GRANUM E., BRYNESTAD S., O'SULLIVAN K. *et al.*, "'Enterotoxin from Bacillus cereus': production and biochemical characterization", *Ned. Melk En Zuiveltijdschr.*, vol. 47, no. 2, pp. 63–70, 1993.

[GRA 02] GRAM L., RAVN L., RASCH M. *et al.*, "Food spoilage—interactions between food spoilage bacteria", *Int. J. Food Microbiol.*, vol. 78, no. 1, pp. 79–97, 2002.

[GRA 04] GRASS G. *et al.*, "Camelysin is a novel surface metalloproteinase from Bacillus cereus", *Infect. Immun.*, vol. 72, no. 1, pp. 219–228, 2004.

[GRI 81] GRIFFITHS M.M., PHILLIPS J., MUIR D., "Thermostability of proteases and lipases from a number of species of psychrotrophic bacteria of dairy origin", *J. Appl. Microbiol.*, vol. 50, no. 2, pp. 289–303, 1981.

[GRI 01] GRIJSPEERDT K., "Modelling the penetration and growth of bacteria in eggs", *Food Control*, vol. 12, no. 1, pp. 7–11, 2001.

[GUE 07] GUERIN-DUBIARD C., NAU F., "Minor Proteins", in HUOPALAHTI P.R., LOPEZ-FANDIÑO R., ANTON M. *et al.* (eds), *Bioactive Egg Compounds*, Springer, Berlin Heidelberg, pp. 93–98, 2007.

[GUE 10] GUERIN-DUBIARD C., ANTON M., GAUTRON J. *et al.*, "Composition de l'œuf", in NAU F., GUERIN-DUBIARD C., BARON F. *et al.* (eds), *Science et technologie de l'oeuf. Vol. 2. De l'œuf aux ovoproduits.*, Tec et Doc, Paris, 2010.

[GUI 03] GUINEBRETIERE M., SANCHIS V., "Bacillus cereus sensu lato", *Bull. Société Française de la Microbiol.*, vol. 18, no. 2, 2003.

[GUI 08] GUINEBRETIÈRE M.-H. *et al.*, "Ecological diversification in the Bacillus cereus group", *Environ. Microbiol.*, vol. 10, no. 4, pp. 851–865, 2008.

[GUN 44] GUNSALUS I., CAMPBELL J., "Diversion of the lactic acid fermentation with oxidized substrate", *J. Bacteriol.*, vol. 48, no. 4, p. 455, 1944.

[GUN 11] GUNCHEVA M., ZHIRYAKOVA D., "Catalytic properties and potential applications of Bacillus lipases", *J. Mol. Catal. B Enzym.*, vol. 68, no. 1, pp. 1–21, January 2011.

[GUP 02] GUPTA R., BEG Q., LORENZ P., "Bacterial alkaline proteases: molecular approaches and industrial applications", *Appl. Microbiol. Biotechnol.*, vol. 59, no. 1, pp. 15–32, 2002.

[HAI 38] HAINES R., "Observations on the bacterial flora of the hen's egg, with a description of new species of Proteus and Pseudomonas causing rots in eggs", *J. Hyg. (Lond.)*, vol. 38, no. 3, p. 338, 1938.

[HAM 97] HAMAD S., DIENG M., EHRMANN M. *et al.*, "Characterization of the bacterial flora of Sudanese sorghum flour and sorghum sourdough", *J. Appl. Microbiol.*, vol. 83, no. 6, pp. 764–770, 1997.

[HAR 63] HARRY E., "The relationship between egg spoilage and the environment of the egg when laid", *Br. Poult. Sci.*, vol. 4, no. 1, pp. 91–100, 1963.

[HEA 76] HEANEY R.K., CURTIS R.F., "Some further studies on incubator reject eggs", *J. Sci. Food Agric.*, vol. 27, no. 11, pp. 1057–1063, 1976.

[HEG 90] HEGAZI F.Z., "Growth rate, proteolysis and acid production of Streptococcus faecalis subsp. liquefaciens in skim milk with some additives", *Mol. Nutr. Food Res.*, vol. 34, no. 2, pp. 195–199, 1990.

[HEM 98] HEMATI B., SUSSMUTH R., EZZAT N. *et al.*, "Isolation and characterization of an Enterococcus starter from Egyptian Domiati cheese", *Milchwissenschaft*, vol. 53, no. 4, pp. 198–202, 1998.

[HER 95] HERGENROTHER P.J., SPALLER M.R., HAAS M.K. *et al.*, "Chromogenic assay for phospholipase C from Bacillus cereus", *Anal. Biochem.*, vol. 229, no. 2, pp. 313–316, 1995.

[HID 95] HIDALGO A., ROSSI M., POMPEI C., "Furosine as a freshness parameter of shell eggs", *J. Agric. Food Chem.*, vol. 43, no. 6, pp. 1673–1677, 1995.

[HID 96] HIDALGO A., LUCISANO M., COMELLI E.M. *et al.*, "Evolution of chemical and physical yolk characteristics during the storage of shell eggs", *J. Agric. Food Chem.*, vol. 44, no. 6, pp. 1447–1452, 1996.

[HID 04] HIDALGO A., ROSSI M., POMPEI C. *et al.*, "Uracil as an Index of Hygienic Quality in Egg Products", *Ital. J. Food Sci.*, vol. 16, no. 4, 2004.

[HID 06] HIDALGO A., ROSSI M., POMPEI C., "Estimation of equivalent egg age through furosine analysis", *Food Chem.*, vol. 94, no. 4, pp. 608–612, 2006.

[HID 08] HIDALGO A., FRANZETTI L., ROSSI M. *et al.*, "Chemical Markers for the Evaluation of Raw Material Hygienic Quality in Egg Products", *J. Agric. Food Chem.*, vol. 56, no. 4, pp. 1289–1297, 2008.

[HIL 04] HILL K.K. *et al.*, "Fluorescent amplified fragment length polymorphism analysis of Bacillus anthracis, Bacillus cereus, and Bacillus thuringiensis isolates", *Appl. Environ. Microbiol.*, vol. 70, no. 2, pp. 1068–1080, 2004.

[HON 98] HONDAL R.J. *et al.*, "Mechanism of phosphatidylinositol-specific phospholipase C: a unified view of the mechanism of catalysis", *Biochemistry (Mosc.)*, vol. 37, no. 13, pp. 4568–4580, 1998.

[HOR 17] HORIMOTO Y., HATTA H., "Egg composition and chemistry. The nutritional and physiological functions of egg yolk components", in ROBERTS J. (ed.), *Achieving Sustainable Production of Eggs, volume. 1: Safety and quality*, Burleigh Dodds, Cambridge, pp. 49–96, 2017.

[HUA 12] HUANG C.-J., TSAY J.-F., CHANG S.-Y. *et al.*, "Dimethyl disulfide is an induced systemic resistance elicitor produced by Bacillus cereus C1L", *Pest Manag. Sci.*, vol. 68, no. 9, pp. 1306–1310, 2012.

[HUG 93] HUGENHOLTZ J., PERDON L., ABEE T., "Growth and energy generation by Lactococcus lactis subsp. lactis biovar diacetylactis during citrate metabolism", *Appl. Environ. Microbiol.*, vol. 59, no. 12, pp. 4216–4222, 1993.

[HUY 91] HUYCKE M.M., SPIEGEL C.A., GILMORE M.S., "Bacteremia caused by hemolytic, high-level gentamicin-resistant Enterococcus faecalis", *Antimicrob. Agents Chemother.*, vol. 35, no. 8, pp. 1626–1634, 1991.

[INO 06] INOUE T., TOMITA H., IKE Y., "Bac 32, a Novel Bacteriocin Widely Disseminated among Clinical Isolates of Enterococcus faecium", *Antimicrob. Agents Chemother.*, vol. 50, no. 4, pp. 1202–1212, April 2006.

[INT 96] IN'T VELD J.H.H., "Microbial and biochemical spoilage of foods: an overview", *Int. J. Food Microbiol.*, vol. 33, no. 1, pp. 1–18, 1996.

[IVA 99] IVANOV I., BOYTCHEVA S., MIHAILOVA G., "Parallel study of thermal resistance and permeability barrier stability of Enterococcus faecalis as affected by salt composition, growth temperature and pre-incubation temperature", *J. Therm. Biol.*, vol. 24, no. 4, pp. 217–227, 1999.

[IVA 03] IVANOVA N. *et al.*, "Genome sequence of Bacillus cereus and comparative analysis with Bacillus anthracis", *Nature*, vol. 423, no. 6935, pp. 87–91, 2003.

[JAE 94] JAEGER K.-E., RANSAC S., DIJKSTRA B.W. *et al.*, "Bacterial lipases", *FEMS Microbiol. Rev.*, vol. 15, no. 1, pp. 29–63, 1994.

[JAN 11] JAN S., BRUNET N., TECHER C. *et al.*, "Biodiversity of psychrotrophic bacteria of the *Bacillus cereus* group collected on farm and in egg product industry", *Food Microbiol.*, vol. 28, no. 2, pp. 261–265, 2011.

[JAY 03] JAY J., VILAI J., HUGHES M., "Profile and activity of the bacterial biota of ground beef held from freshness to spoilage at 5–7°C", *Int. J. Food Microbiol.*, vol. 81, no. 2, pp. 105–111, 2003.

[JEN 75a] JENSEN J., REINBOLD G., WASHAM C. *et al.*, "Role of enterococci in Cheddar cheese: free fatty acid appearance and citric acid utilization", *J. Milk Food Technol.*, vol. 38, no. 2, pp. 78–83, 1975.

[JEN 75b] JENSEN J.P., REINBOLD G., WASHAM C. *et al.*, "Role of enterococci in Cheddar cheese: proteolytic activity and lactic acid development", *J. Milk Food Technol.*, vol. 38, no. 1, pp. 3–7, 1975.

[JET 94] JETT B.D., HUYCKE M.M., GILMORE M.S., "Virulence of enterococci", *Clin. Microbiol. Rev.*, vol. 7, no. 4, pp. 462–478, 1994.

[JON 79] JONES J., ELLINGWORTH C.E., "Detection of incubator rejects in whole eggs after normal commercial processing", *Int. J. Food Sci. Technol.*, vol. 14, no. 2, pp. 199–203, 1979.

[KAL 92] KALOGRIDOU-VASSILIADOU D., "Biochemical activities of *Bacillus* species isolated from flat sour evaporated milk", *J. Dairy Sci.*, vol. 75, no. 10, pp. 2681–2686, 1992.

[KAN 07] KANKI M., YODA T., TSUKAMOTO T. *et al.*, "Histidine decarboxylases and their role in accumulation of histamine in tuna and dried saury", *Appl. Environ. Microbiol.*, vol. 73, no. 5, pp. 1467–1473, 2007.

[KAR 05] KAROVIČOVÁ J., KOHAJDOVA Z., "Biogenic amines in food", *Chem. Pap.,* vol. 59, no. 1, pp. 70–79, 2005.

[KAU 15] KAUR B., KAUR R., "Isolation, identification and genetic organization of the ADI operon in *Enterococcus faecium* GR7", *Ann. Microbiol.*, vol. 65, no. 3, pp. 1427–1437, 2015.

[KAY 03] KAYSER F.H., "Safety aspects of enterococci from the medical point of view", *Int. J. Food Microbiol.*, vol. 88, nos 2–3, pp. 255–262, 2003.

[KAY 04] KAYAOGLU G., ØRSTAVIK D., "Virulence factors of *Enterococcus faecalis*: relationship to endodontic disease", *Crit. Rev. Oral Biol. Med.*, vol. 15, no. 5, pp. 308–320, 2004.

[KIM 02] KIM H.K., CHOI J., KIM M.H. *et al.*, "Expression and characterization of Ca^{2+}-independent lipase from *Bacillus pumilus* B26", *Biochim. Biophys. Acta BBA-Mol. Cell Biol. Lipids*, vol. 1583, no. 2, pp. 205–212, 2002.

[KIM 08] KIM Y.-R., BATT C.A., "Riboprint and virulence gene patterns for *Bacillus cereus* and related species", *J. Microbio. Biotechnol.,* vol. 18, no. 6, pp. 1146–1155, 2008.

[KLE 01] KLEEREBEZEM M., QUADRI L.E., "Peptide pheromone-dependent regulation of antimicrobial peptide production in Gram-positive bacteria: a case of multicellular behavior", *Peptides*, vol. 22, no. 10, pp. 1579–1596, 2001.

[KLE 03] KLEIN G., "Taxonomy, ecology and antibiotic resistance of enterococci from food and the gastro-intestinal tract", *Int. J. Food Microbiol.*, vol. 88, no. 2, pp. 123–131, 2003.

[KNU 93] KNUDTSON L.M., HARTMAN P.A., "Enterococci in pork processing", *J. Food Prot.*, vol. 56, no. 1, pp. 6–9, 1993.

[KOC 04] KOCH S., HUFNAGEL M., THEILACKER C. *et al.*, "Enterococcal infections: host response, therapeutic, and prophylactic possibilities", *Vaccine*, vol. 22, no. 7, pp. 822–830, 2004.

[KOM 65] KOMINEK L.A., HALVORSON H.O., "Metabolism of poly-β-hydroxybutyrate and acetoin in *Bacillus cereus*", *J. Bacteriol.*, vol. 90, no. 5, pp. 1251–1259, 1965.

[KOM 08] KOMPRDA T., BURDYCHOVÁ R., DOHNAL V. *et al.*, "Tyramine production in Dutch-type semi-hard cheese from two different producers", *Food Microbiol.*, vol. 25, no. 2, pp. 219–227, 2008.

[KON 13] KONE A.Z., JAN S., Le MARECHAL C. *et al.*, "Identifying risk factors for eggshell contamination by *Bacillus cereus* group bacteria in French laying farms", *Br. Poult. Sci.*, pp. 1–8, 2013.

[KOU 10] KOUTSOUMANIS K., TASSOU C., NYCHAS G.-J.E., "Biogenic Amines in Foods", in JUNEJA V.K., SOFOS J.N. (eds), *Pathogens and Toxins in Foods*, American Society of Microbiology, Washington, 2010.

[KUP 89] KUPPE A., EVANS L.M., MCMILLEN D.A. *et al.*, "Phosphatidylinositol-specific phospholipase C of *Bacillus cereus*: cloning, sequencing, and relationship to other phospholipases", *J. Bacteriol.*, vol. 171, no. 11, pp. 6077–6083, 1989.

[KUS 57] KUSHNER D.J., "An evaluation of the egg-yolk reaction as a test for lecithinase activity", *J. Bacteriol.*, vol. 73, no. 3, pp. 297–302, 1957.

[LAD 11a] LADERO V., RATTRAY F.P., MAYO B. *et al.*, "Sequencing and transcriptional analysis of the biosynthesis gene cluster of putrescine-producing *Lactococcus lactis*", *Appl. Environ. Microbiol.*, vol. 77, no. 18, pp. 6409–6418, 2011.

[LAD 11b] LADERO V., SÁNCHEZ-LLANA E., FERNÁNDEZ M. *et al.*, "Survival of biogenic amine-producing dairy LAB strains at pasteurisation conditions", *Int. J. Food Sci. Technol.*, vol. 46, no. 3, pp. 516–521, 2011.

[LAD 12] LADERO V., FERNÁNDEZ M., CALLES-ENRÍQUEZ M. *et al.*, "Is the production of the biogenic amines tyramine and putrescine a species-level trait in enterococci?", *Food Microbiol.*, vol. 30, no. 1, pp. 132–138, 2012.

[LAN 33] LANCEFIELD R.C., "A serological differentiation of human and other groups of hemolytic streptococci", *J. Exp. Med.*, vol. 57, no. 4, pp. 571–595, 1933.

[LAO 14] LAOUAMI S., CLAIR G., ARMENGAUD J. *et al.*, "Proteomic evidences for rex regulation of metabolism in toxin-producing *Bacillus cereus* ATCC 14579", *PLoS ONE*, vol. 9, no. 9, p. e107354, 2014.

[LAT 11] LATORRE-MORATALLA M., BOSCH-FUSTÉ J., BOVER-CID S. *et al.*, "Contribution of enterococci to the volatile profile of slightly-fermented sausages", *LWT-Food Sci. Technol.*, vol. 44, no. 1, pp. 145–152, 2011.

[LEB 06] LEBLANC D.J., "Enterococcus", in Dworkin M., Falkow S., ROSENBERG E. *et al.* (eds), *The Prokaryotes*, vol. 4, 3rd ed., Springer, New York, 2006.

[LEB 14] LEBRETON F., WILLEMS R.J., GILMORE M.S., "Enterococcus diversity, origins in nature, and gut colonization", in GILMORE S., CLEWELL D.B., IKE Y. *et al.* (eds), *Enterococci: From Commensals to Leading Causes of Drug Resistant Infection*, Massachusetts Eye and Ear Infirmary, Boston, 2014.

[LED 09] LEDENBACH L.H., MARSHALL R.T., "Microbiological spoilage of dairy products", in SPERBER W.H., DOYLE M.P. (eds), *Compendium of the Microbiological Spoilage of Foods and Beverages*, Springer, New York, pp. 41–67, 2009.

[LER 96] LERECLUS D., AGAISSE H., GOMINET M. *et al.*, "Identification of a *Bacillus thuringiensis* gene that positively regulates transcription of the phosphatidylinositol-specific phospholipase C gene at the onset of the stationary phase", *J. Bacteriol.*, vol. 178, no. 10, pp. 2749–2756, 1996.

[LER 00] LERECLUS D., AGAISSE H., GRANDVALET C. *et al.*, "Regulation of toxin and virulence gene transcription in *Bacillus thuringiensis*", *Int. J. Med. Microbiol.*, vol. 290, nos 4–5, pp. 295–299, 2000.

[LIA 89] LIAO C.-H., "Analysis of pectate lyases produced by soft rot bacteria associated with spoilage of vegetables", *Appl. Environ. Microbiol.*, vol. 55, no. 7, pp. 1677–1683, 1989.

[LIN 98] LINDBERG A.-M., LJUNGH A.A., AHRNE S. *et al.*, "Enterobacteriaceae found in high numbers in fish, minced meat and pasteurised milk or cream and the presence of toxin encoding genes", *Int. J. Food Microbiol.*, vol. 39, no. 1, pp. 11–17, 1998.

[LIN 00] LINDSAY D., BRÖZEL V., MOSTERT J. *et al.*, "Physiology of dairy-associated *Bacillus* spp. over a wide pH range", *Int. J. Food Microbiol.*, vol. 54, no. 1, pp. 49–62, 2000.

[LIN 11] LINARES D.M., MARTÍN M., LADERO V. *et al.*, "Biogenic amines in dairy products", *Crit. Rev. Food Sci. Nutr.*, vol. 51, no. 7, pp. 691–703, 2011.

[LIU 03] LIU M., GRIFFITHS M., "The role of quorum sensing in the regulation of protease by *Pseudomonas fluorescens* and its relation to food spoilage", *IFT Annual Meeting, vol. Book of Abstracts, 29G-26,* Chicago, p. 85, 2003.

[LIU 06] LIU M., GRAY J., GRIFFITHS M., "Occurrence of proteolytic activity and N-acyl-homoserine lactone signals in the spoilage of aerobically chill-stored proteinaceous raw foods", *J. Food Prot.*, vol. 69, no. 11, pp. 2729–2737, 2006.

[LIU 07] LIU M., WANG H., GRIFFITHS M., "Regulation of alkaline metalloprotease promoter by N-acyl homoserine lactone quorum sensing in *Pseudomonas fluorescens*", *J. Appl. Microbiol.*, vol. 103, no. 6, pp. 2174–2184, 2007.

[LIU 14] LIU F., DU L., WANG D. *et al.*, "Heterologous expression and characterization of tyrosine decarboxylase from Enterococcus faecalis R612Z1 and *Enterococcus faecium* R615Z1", *J. Food Prot.*, vol. 77, no. 4, pp. 592–598, 2014.

[LU 04] LU L., HUME M.E., PILLAI S.D., "Autoinducer-2–like activity associated with foods and its interaction with food additives", *J. Food Prot.*, vol. 67, no. 7, pp. 1457–1462, 2004.

[LUC 96] LUCISANO M., HIDALGO A., COMELLI E.M. *et al.*, "Evolution of chemical and physical albumen characteristics during the storage of shell eggs", *J. Agric. Food Chem.*, vol. 44, no. 5, pp. 1235–1240, 1996.

[LUN 65] LUND B.M., "A comparison by the use of gel electrophoresis of soluble protein components and esterase enzymes of some group D streptococci", *Microbiology*, vol. 40, no. 3, pp. 413–419, 1965.

[LUN 97] LUND T., GRANUM P.E., "Comparison of biological effect of the two different enterotoxin complexes isolated from three different strains of *Bacillus cerous*", *Microbiology*, vol. 143, no. 10, pp. 3329–3336, 1997.

[LUO 07] LUO B., GROENKE K., TAKORS R. *et al.*, "Simultaneous determination of multiple intracellular metabolites in glycolysis, pentose phosphate pathway and tricarboxylic acid cycle by liquid chromatography–mass spectrometry", *J. Chromatogr. A*, vol. 1147, no. 2, pp. 153–164, 2007.

[LYO 04] LYON G.J., NOVICK R.P., "Peptide signaling in *Staphylococcus aureus* and other Gram-positive bacteria", *Peptides*, vol. 25, no. 9, pp. 1389–1403, 2004.

[MAC 41] MACFARLANE R., OAKLEY C., ANDERSON C., "Hemolysis and the production of opalescence in serum and lecitho-vitellin by the α toxin of clostridium welchii", *J. Pathol.*, vol. 52, no. 1, pp. 99–103, 1941.

[MAC 82] MACKENZIE K., SKERMAN V., "Microbial spoilage in unpasteurised liquid whole egg", *Food Technol. Aust.*, 1982.

[MAC 94] MACIVER B., MCHALE R.H., SAUL D.J. *et al.*, "Cloning and sequencing of a serine proteinase gene from a thermophilic *Bacillus* species and its expression in *Escherichia coli*", *Appl. Environ. Microbiol.*, vol. 60, no. 11, pp. 3981–3988, 1994.

[MAC 97] MACEDO A.C., MALCATA F.X., "Role of adventitious microflora in proteolysis and lipolysis of Serra cheese: preliminary screening", *Z. Für Leb. -Forsch. A*, vol. 205, no. 1, pp. 25–30, 1997.

[MAC 17] MACHADO S.G. *et al.*, "The biodiversity of the microbiota producing heat-resistant enzymes responsible for spoilage in processed bovine milk and dairy products", *Front. Microbiol.*, vol. 8, 2017.

[MAG 86] MAGNUS C.C., INGLEDEW W.W., MCCURDY A., "Thermal resistance of streptococci isolated from pasteurized ham", *Can. Inst. Food Sci. Technol. J.*, vol. 19, no. 2, pp. 62–67, 1986.

[MAG 88] MAGNUS C., MCCURDY A., INGLEDEW W., "Further studies on the thermal resistance of *Streptococcus faecium* and *Streptococcus faecalis* in pasteurized ham", *Can. Inst. Food Sci. Technol. J.*, vol. 21, no. 2, pp. 209–212, 1988.

[MAJ 17] MAJUMDER K., MINE Y., "Egg composition and chemistry. Composition and properties of egg white", *Achieving Sustainable Production of Eggs Volume 1: Safety and Quality*, Burleigh Dodds Series in Agricultural Science, Cambridge, pp. 25–47, 2017.

[MAR 76] MARTINEZ-MORENO J., "Microbial flora of Manchego cheese: III. Streptococci", *An. Inst. Nac. Investig. Agrar. Ser. Gen.*, vol. 4, pp. 41–56, 1976.

[MAR 00] MARTIN S.F., FOLLOWS B.C., HERGENROTHER P.J. *et al.*, "The choline binding site of phospholipase C (*Bacillus cereus*): insights into substrate specificity", *Biochemistry (Mosc.)*, vol. 39, no. 12, pp. 3410–3415, 2000.

[MAR 03] MARTINEZ S., LOPEZ M., BERNARDO A., "Thermal inactivation of *Enterococcus faecium*: effect of growth temperature and physiological state of microbial cells", *Lett. Appl. Microbiol.*, vol. 37, no. 6, pp. 475–481, 2003.

[MAR 04a] MARCH J.C., BENTLEY W.E., "Quorum sensing and bacterial cross-talk in biotechnology", *Curr. Opin. Biotechnol.*, vol. 15, no. 5, pp. 495–502, 2004.

[MAR 04b] MARCIŇÁKOVÁ M., SIMONOVÁ M., LAUKOVÁ A., "Probiotic properties of *Enterococcus faecium* EF9296 strain isolated from silage", *Acta Vet. Brno*, vol. 73, no. 4, pp. 513–519, 2004.

[MAR 06] MARCOBAL A., DE LAS RIVAS B., MUÑOZ R., "First genetic characterization of a bacterial β-phenylethylamine biosynthetic enzyme in *Enterococcus faecium* RM58", *FEMS Microbiol. Lett.*, vol. 258, no. 1, pp. 144–149, 2006.

[MAR 09] MARTÍNEZ-BLANCH J.F., SÁNCHEZ G., GARAY E., *et al.*, "Development of a real-time PCR assay for detection and quantification of enterotoxigenic members of *Bacillus cereus* group in food samples", *Int. J. Food Microbiol.*, vol. 135, no. 1, pp. 15–21, September 2009.

[MAR 12] MARCOBAL A., DE LAS RIVAS B., LANDETE J.M. *et al.*, "Tyramine and phenylethylamine biosynthesis by food bacteria", *Crit. Rev. Food Sci. Nutr.*, vol. 52, no. 5, pp. 448–467, 2012.

[MAY 96] MAYR B., KAPLAN T., LECHNER S. *et al.*, "Identification and purification of a family of dimeric major cold shock protein homologs from the psychrotrophic *Bacillus cereus* WSBC 10201", *J. Bacteriol.*, vol. 178, no. 10, pp. 2916–2925, 1996.

[MCG 48] MCGAUGHEY C., CHU H., "The egg-yolk reaction of aerobic sporing bacilli", *Microbiology*, vol. 2, no. 3, pp. 334–340, 1948.

[MCK 84] MCKELLAR R.C., NICHOLS-NELSON D., "Acid-coagulation of evaporated milk by a coculture of *Enterococcus faecium* and *Bacillus subtilis*", *J. Food Prot.*, vol. 47, no. 11, pp. 853–855, 1984.

[MCS 00] MCSWEENEY P.L., SOUSA M.J., "Biochemical pathways for the production of flavour compounds in cheeses during ripening: a review", *Le Lait*, vol. 80, no. 3, pp. 293–324, 2000.

[MEE 91] MEER R., BAKER J., BODYFELT F. *et al.*, "Psychrotrophic *Bacillus* spp. in fluid milk products: a review", *J. Food Prot.*, vol. 54, no. 12, pp. 969–979, 1991.

[MIC 03] MICHELET N., MAHILLON J., "*Bacillus cereus* : opportuniste et pathogène", *Bulletin de la société française de microbiologie,* vol. 18, pp. 113–122, 2003.

[MIL 10] MILLER P. *et al.*, "Shelf life extension of liquid whole eggs by heat and bacteriocin treatment", *Czech J. Food Sci.*, vol. 28, no. 4, pp. 280–289, 2010.

[MIR 07] MIRANDA J.M. *et al.*, "Antimicrobial resistance in *Enterococcus* spp. strains isolated from organic chicken, conventional chicken, and turkey meat: a comparative survey", *J. Food Prot.*, vol. 70, no. 4, pp. 1021–1024, 2007.

[MIY 00] MIYOSHI S., SHINODA S., "Microbial metalloproteases and pathogenesis", *Microbes Infect.*, vol. 2, no. 1, pp. 91–98, 2000.

[MOL 07] MOLS M., DE BEEN M., ZWIETERING M.H. *et al.*, "Metabolic capacity of *Bacillus cereus* strains ATCC 14579 and ATCC 10987 interlinked with comparative genomics", *Environ. Microbiol.*, vol. 9, no. 12, pp. 2933–2944, 2007.

[MOL 11] MOLS M., ABEE T., "Primary and secondary oxidative stress in *Bacillus*", *Environ. Microbiol.*, vol. 13, no. 6, pp. 1387–1394, 2011.

[MON 98] MONTEL M., MASSON F., TALON R., "Bacterial role in flavour development", *Meat Sci.*, vol. 49, pp. S111–S123, 1998.

[MOR 87] MORRIS C.E., "Determination of uracil, uridine and formic acid in egg products by high-performance liquid chromatography", *J. Chromatogr. A*, vol. 394, no. 2, pp. 408–413, 1987.

[MOR 04] MORIARTY-CRAIGE S.E., JONES D.P., "Extracellular thiols and thiol/disulfide redox in metabolism", *Annu. Rev. Nutr.*, vol. 24, no. 1, pp. 481–509, 2004.

[MOR 06] MORENO M.F., SARANTINOPOULOS P., TSAKALIDOU E. *et al.*, "The role and application of enterococci in food and health", *Int. J. Food Microbiol.*, vol. 106, no. 1, pp. 1–24, 2006.

[MOS 95] MOSSO C., RIGO S., DI PIETRO S., "Acidi lattico e 3 OH-butirrico in uova di scarto di incubatoio: considerazioni sulla determinazione analitica spettrofotometrica", *Ind. Aliment.*, vol. 34, no. 335, pp. 259–260, 1995.

[MUL 90] MULYOWIDARSO R., FLEET G., BUCKLE K., "Association of bacteria with the fungal fermentation of soybean tempe", *J. Appl. Microbiol.*, vol. 68, no. 1, pp. 43–47, 1990.

[MUS 05] MUSGROVE M.T., JONES D.R., NORTHCUTT J.K. *et al.*, "Impact of commercial processing on the microbiology of shell eggs", *J. Food Prot.*, vol. 68, no. 11, pp. 2367–2375, 2005.

[NAB 10] NABRDALIK M., GRATA K., LATALA A., "Proteolytic activity of *Bacillus cereus* strains", *Proc. ECOpole*, vol. 4, no. 2, pp. 273–277, 2010.

[NEE 31] NEEDHAM J., "The relations between yolk and white in the hen's egg", *J. Exp. Biol.*, vol. 8, no. 3, pp. 330–344, 1931.

[NIE 11] NIETO-ARRIBAS P., SESEÑA S., POVEDA J.M. *et al.*, "Enterococcus populations in artisanal Manchego cheese: Biodiversity, technological and safety aspects", *Food Microbiol.*, vol. 28, no. 5, pp. 891–899, 2011.

[NOU 94] NOUT M., "Fermented foods and food safety", *Food Res. Int.*, vol. 27, no. 3, pp. 291–298, 1994.

[NYC 08] NYCHAS G.-J.E., SKANDAMIS P.N., TASSOU C.C. *et al.*, "Meat spoilage during distribution", *Meat Sci.*, vol. 78, no. 1, pp. 77–89, 2008.

[OGI 08] OGIER J-C., SERROR P., "Safety assessment of dairy microorganisms: the *Enterococcus* genus", *Int. J. Food Microbiol.*, vol. 126, no. 3, pp. 291–301, 2008.

[ORD 78] ORDONEZ J., BARNETO R., RAMOS M., "Studies on Manchego cheese ripened in olive oil", *Milchwissenschaft*, vol. 33, no. 10, pp. 609–613, 1978.

[OUH 06] OUHIB O., CLAVEL T., SCHMITT P., "The production of *Bacillus cereus* enterotoxins is influenced by carbohydrate and growth rate", *Curr. Microbiol.*, vol. 53, no. 3, pp. 222–226, 2006.

[OUH 09] OUHIB-JACOBS O., LINDLEY N.D., SCHMITT P. *et al.*, "Fructose and glucose mediates enterotoxin production and anaerobic metabolism of *Bacillus cereus* ATCC14579 ^{T}", *J. Appl. Microbiol.*, vol. 107, no. 3, pp. 821–829, 2009.

[OUM 01] OUMER A. *et al.*, "Proteolysis and formation of volatile compounds in cheese manufactured with a bacteriocin-producing adjunct culture", *J. Dairy Res.*, vol. 68, no. 1, pp. 117–129, 2001.

[PAN 10] PANDIANI F. *et al.*, "Differential involvement of the five RNA helicases in adaptation of *Bacillus cereus* ATCC 14579 to low growth temperatures", *Appl. Environ. Microbiol.*, vol. 76, no. 19, pp. 6692–6697, 2010.

[PAN 11] PANDIANI F., CHAMOT S., BRILLARD J. *et al.*, "Role of the five RNA helicases in the adaptive response of *Bacillus cereus* ATCC 14579 cells to temperature, pH, and oxidative stresses", *Appl. Environ. Microbiol.*, vol. 77, no. 16, pp. 5604–5609, 2011.

[PAP 16] PAPENFORT K.K., BASSLER B.L., "Quorum sensing signal-response systems in Gram-negative bacteria", *Nat. Rev. Microbiol.*, vol. 14, no. 9, pp. 576–588, 2016.

[PAY 79] PAYNE J., GOOCH J., BARNES E.M., "Heat-resistant bacteria in pasteurized whole egg", *J. Appl. Microbiol.*, vol. 46, no. 3, pp. 601–613, 1979.

[PER 82] PEREZ S.,. LORENZO P.L., GARCIA M. *et al.*, "Heat resistance of enterococci", *Milchwissenschaft*, vol. 37, no. 12, pp. 724–726, 1982.

[PER 09] PEREIRA C., MATOS D., SAN ROMÃO M. *et al.*, "Dual role for the tyrosine decarboxylation pathway in *Enterococcus faecium* E17: response to an acid challenge and generation of a proton motive force", *Appl. Environ. Microbiol.*, vol. 75, no. 2, pp. 345–352, 2009.

[PER 11] PERCHAT S. *et al.*, "A cell–cell communication system regulates protease production during sporulation in bacteria of the *Bacillus cereus* group", *Mol. Microbiol.*, vol. 82, no. 3, pp. 619–633, 2011.

[PER 15] PEREZ M. *et al.*, "Tyramine biosynthesis is transcriptionally induced at low pH and improves the fitness of *Enterococcus faecalis* in acidic environments", *Appl. Microbiol. Biotechnol.*, vol. 99, no. 8, pp. 3547–3558, 2015.

[PER 16] PEREZ-PASCUAL D., MONNET V., GARDAN R., "Bacterial cell–cell communication in the host via RRNPP peptide-binding regulators", *Front. Microbiol.*, vol. 7, 2016.

[PET 03] PETERS J., MAC K., WICHMANN-SCHAUER H. *et al.*, "Species distribution and antibiotic resistance patterns of enterococci isolated from food of animal origin in Germany", *Int. J. Food Microbiol.*, vol. 88, no. 2, pp. 311–314, 2003.

[PIN 07] PINTO U.M., DE SOUZA VIANA E., MARTINS M.L. *et al.*, "Detection of acylated homoserine lactones in Gram-negative proteolytic psychrotrophic bacteria isolated from cooled raw milk", *Food Control*, vol. 18, no. 10, pp. 1322–1327, 2007.

[PIR 93] PIRHONEN M., FLEGO D., HEIKINHEIMO R. *et al.*, "A small diffusible signal molecule is responsible for the global control of virulence and exoenzyme production in the plant pathogen *Erwinia carotovora*", *EMBO J.*, vol. 12, no. 6, p. 2467, 1993.

[PIR 96] PIRTTIJÄRVI T., GRAEFFE T., SALKINOJA-SALONEN M., "Bacterial contaminants in liquid packaging boards: assessment of potential for food spoilage", *J. Appl. Microbiol.*, vol. 81, no. 4, pp. 445–458, 1996.

[PIR 00] PIRTTIJÄRVI T., ANDERSSON M., SALKINOJA-SALONEN M., "Properties of *Bacillus cereus* and other bacilli contaminating biomaterial-based industrial processes", *Int. J. Food Microbiol.*, vol. 60, no. 2, pp. 231–239, 2000.

[POD 04] PODBIELSKI A., KREIKEMEYER B., "Cell density–dependent regulation: basic principles and effects on the virulence of Gram-positive cocci", *Int. J. Infect. Dis.*, vol. 8, no. 2, pp. 81–95, 2004.

[POS 10] POSTOLLEC F. *et al.*, "A multiparametric PCR-based tool for fast detection and identification of spore-forming bacteria in food", *Int. J. Food Microbiol.*, vol. 142, no. 1, pp. 78–88, 2010.

[PRI 77] PRIEST F.G., "Extracellular enzyme synthesis in the genus *Bacillus*", *Bacteriol. Rev.*, vol. 41, no. 3, p. 711, 1977.

[PRO 07] PROTAIS J., BOSCHER E., QUEGUINER S. *et al.*, "Identification of the *Enterococcus* genus in the matrix of liquid whole eggs", *7èmes Journ. Rech. Avic. Tours France,* pp. 528–531, 2007.

[PRÜ 99a] PRÜSS B.M., DIETRICH R., NIBLER B. *et al.*, "The hemolytic enterotoxin HBL is broadly distributed among species of the *Bacillus cereus* group", *Appl. Environ. Microbiol.*, vol. 65, no. 12, pp. 5436–5442, 1999.

[PRÜ 99b] PRÜSS B.M., FRANCIS K.P., VON STETTEN F. *et al.*, "Correlation of 16S ribosomal DNA signature sequences with temperature-dependent growth rates of mesophilic and psychrotolerant strains of the *Bacillus cereus* group", *J. Bacteriol.*, vol. 181, no. 8, pp. 2624–2630, 1999.

[QIN 00] QIN X., SINGH K.V., WEINSTOCK G.M. *et al.*, “Effects of *Enterococcus faecalis* fsr genes on production of gelatinase and a serine protease and virulence”, *Infect. Immun.*, vol. 68, no. 5, pp. 2579–2586, 2000.

[RAJ 02] RAJMOHAN S., DODD C., WAITES W., “Enzymes from isolates of *Pseudomonas fluorescens* involved in food spoilage”, *J. Appl. Microbiol.*, vol. 93, no. 2, pp. 205–213, 2002.

[RAM 05] RAMARAO N., LERECLUS D., “The InhA1 metalloprotease allows spores of the *B. cereus* group to escape macrophages: InhA1 allows *B. thuringiensis* escape from macrophages”, *Cell. Microbiol.*, vol. 7, no. 9, pp. 1357–1364, 2005.

[RAN 07] RANDAZZO C.L. *et al.*, “Preliminary characterization of wild lactic acid bacteria and their abilities to produce flavour compounds in ripened model cheese system”, *J. Appl. Microbiol.*, vol. 103, no. 2, pp. 427–435, 2007.

[RAS 05] RASKO D.A., ALTHERR M.R., HAN C.S. *et al.*, “Genomics of the *Bacillus cereus* group of organisms”, *FEMS Microbiol. Rev.*, vol. 29, no. 2, pp. 303–329, 2005.

[REA 03a] REA M.C., COGAN T.M., “Glucose prevents citrate metabolism by enterococci”, *Int. J. Food Microbiol.*, vol. 88, no. 2, pp. 201–206, 2003.

[REA 03b] REA M.C., COGAN T.M., “Catabolite repression in *Enterococcus faecalis*”, *Syst. Appl. Microbiol.*, vol. 26, no. 2, pp. 159–164, 2003.

[REA 07] READING N.C., TORRES A.G., KENDALL M.M. *et al.*, “A novel two-component signaling system that activates transcription of an enterohemorrhagic *Escherichia coli* effector involved in remodeling of host actin”, *J. Bacteriol.*, vol. 189, no. 6, pp. 2468–2476, 2007.

[RIE 01] RIEDEL K. *et al.*, “N-acyl-l-homoserine lactone-mediated regulation of the lip secretion system in *Serratia liquefaciens* MG1”, *J. Bacteriol.*, vol. 183, no. 5, pp. 1805–1809, 2001.

[RIV 01] RIVA M., FESSAS D., SCHIRALDI A., “Isothermal calorimetry approach to evaluate shelf life of foods”, *Thermochim. Acta*, vol. 370, no. 1, pp. 73–81, 2001.

[ROB 75] ROBINSON D.S., BARNES E.M., TAYLOR J., "Occurrence of β-hydroxybutyric acid in incubator reject eggs", *J. Sci. Food Agric.*, vol. 26, no. 1, pp. 91–98, 1975.

[ROC 10] ROCHA-ESTRADA J., ACEVES-DIEZ A.E., GUARNEROS G. *et al.*, "The RNPP family of quorum-sensing proteins in Gram-positive bacteria", *Appl. Microbiol. Biotechnol.*, vol. 87, no. 3, pp. 913–923, 2010.

[ROM 07] ROMANO P., CAPECE A., POETA C., "Biogenic amine formation in alcoholic fermentation", *Bull OIV*, vol. 80, pp. 251–262, 2007.

[ROM 12] ROMANO A., TRIP H., LONVAUD-FUNEL A. *et al.*, "Evidence of two functionally distinct ornithine decarboxylation systems in lactic acid bacteria", *Appl. Environ. Microbiol.*, vol. 78, no. 6, pp. 1953–1961, 2012.

[ROS 95] ROSSI M., HIDALGO A., POMPEI C., "Furosine as a freshness parameter of shell eggs", *J. Agri. Food Chem.*, vol. 43, no. 6, pp. 1673–1677, 1995.

[ROS 99] ROSSI M., HIDALGO A., POMPEI C. *et al.*, "A new approach to the identification of incubator reject eggs in pasteurized whole egg products obtained from fresh eggs", *European Symposium on the Quality of Eggs and Egg Products*, pp. 451–456, 1999.

[ROS 05] ROSENFELD E., DUPORT C., ZIGHA A. *et al.*, "Characterization of aerobic and anaerobic vegetative growth of the food-borne pathogen *Bacillus cereus* F4430/73 strain", *Can. J. Microbiol.*, vol. 51, no. 2, pp. 149–158, 2005.

[ROS 10] ROSSI M., CASIRAGHI E., PRIMAVESI L. *et al.*, "Functional properties of pasteurised liquid whole egg products as affected by the hygienic quality of the raw eggs", *LWT – Food Sci. Technol.*, vol. 43, no. 3, pp. 436–441, 2010.

[RUI 04] RUIZ-CAPILLAS C., JIMÉNEZ-COLMENERO F., "Biogenic amine content in Spanish retail market meat products treated with protective atmosphere and high pressure", *Eur. Food Res. Technol.*, vol. 218, no. 3, pp. 237–241, 2004.

[RUI 05] RUIZ C., PASTOR F.I.J., DIAZ P., "Isolation of lipid- and polysaccharide-degrading micro-organisms from subtropical forest soil, and analysis of lipolytic strain *Bacillus* sp. CR-179", *Lett. Appl. Microbiol.*, vol. 40, no. 3, pp. 218–227, 2005.

[SAB 04] SABIA C., MESSI P., DE NIEDERHAUSERN S., MANICARDI G. *et al.*, "Study of two bacteriocins produced by *Enterococcus casseliflavus* and *Ent. faecalis*", *Lett. Appl. Microbiol.*, vol. 38, no. 2, pp. 99–105, 2004.

[SAD 16] SADIQ F.A. *et al.*, "The heat resistance and spoilage potential of aerobic mesophilic and thermophilic spore forming bacteria isolated from Chinese milk powders", *Int. J. Food Microbiol.*, vol. 238, pp. 193–201, 2016.

[SAL 72] SALWIN H., STARUSZKIEWICZ W.F.J., BOND J.F., "Formation of -hydroxybutyric acid in eggs during incubation", *J. – Assoc. Off. Anal. Chem.*, vol. 55, no. 3, pp. 458–460, 1972.

[SÁN 07] SÁNCHEZ J., BASANTA A., GÓMEZ-SALA B. *et al.*, "Antimicrobial and safety aspects, and biotechnological potential of bacteriocinogenic enterococci isolated from mallard ducks (*Anas platyrhynchos*)", *Int. J. Food Microbiol.*, vol. 117, no. 3, pp. 295–305, 2007.

[SAR 01a] SARANTINOPOULOS P. *et al.*, "Biochemical properties of enterococci relevant to their technological performance", *Int. Dairy J.*, vol. 11, no. 8, pp. 621–647, 2001.

[SAR 01b] SARANTINOPOULOS P., KALANTZOPOULOS G., TSAKALIDOU E., "Citrate metabolism by *Enterococcus faecalis* FAIR-E 229", *Appl. Environ. Microbiol.*, vol. 67, no. 12, pp. 5482–5487, 2001.

[SAR 02] SARANTINOPOULOS P., KALANTZOPOULOS G., TSAKALIDOU E., "Effect of *Enterococcus faecium* on microbiological, physicochemical and sensory characteristics of Greek Feta cheese", *Int. J. Food Microbiol.*, vol. 76, no. 1, pp. 93–105, 2002.

[SAR 17] SARASWAT R., VERMA V., SISTLA S. *et al.*, "Evaluation of alkali and thermotolerant lipase from an indigenous isolated *Bacillus* strain for detergent formulation", *Electron. J. Biotechnol.*, vol. 30, pp. 33–38, 2017.

[SCH 84] SCHLEIFER K.H., KILPPER-BÄLZ R., "Transfer of *Streptococcus faecalis* and *Streptococcus faecium* to the Genus *Enterococcus* nom. rev. as *Enterococcus faecalis* comb. nov. and *Enterococcus faecium* comb. nov", *Int. J. Syst. Evol. Microbiol.*, vol. 34, no. 1, pp. 31–34, 1984.

[SCH 87] SCHLEIFER K.H., KILPPER-BÄLZ R., "Molecular and chemotaxonomic approaches to the classification of streptococci, enterococci and lactococci: a review", *Syst. Appl. Microbiol.*, vol. 10, no. 1, pp. 1–19, 1987.

[SCH 00] SCHULTEN S. *et al.*, "Evaluation of the ISO 7932 standard for the enumeration of *Bacillus cereus* in foods", *Int. J. Food Microbiol.*, vol. 57, no. 1, pp. 53–61, 2000.

[SCH 05] SCHOENI J.L., LEE WONG A.C., "*Bacillus cereus* food poisoning and its toxins", *J. Food Prot.*, vol. 68, no. 3, pp. 636–648, 2005.

[SEM 03] SEMEDO T., SANTOS M.A., LOPES M.F. *et al.*, "Virulence factors in food, clinical and reference enterococci: a common trait in the genus?", *Syst. Appl. Microbiol.*, vol. 26, no. 1, pp. 13–22, 2003.

[SHA 70] SHAFI R., COTTERILL O., NICHOLS M., "Microbial flora of commercially pasteurized egg products", *Poult. Sci.*, vol. 49, no. 2, pp. 578–585, 1970.

[SHA 94] SHAH N., "Psychrotrophs in milk: a review", *Milchwiss. Ger.*, 1994.

[SHA 96] SHALABY A.R., "Significance of biogenic amines to food safety and human health", *Food Res. Int.*, vol. 29, no. 7, pp. 675–690, 1996.

[SHA 98] SHAPIRO J.A., "Thinking about bacterial populations as multicellular organisms", *Annu. Rev. Microbiol.*, vol. 52, no. 1, pp. 81–104, 1998.

[SHE 37] SHERMAN J.M., "The streptococci", *Bacteriol. Rev.*, vol. 1, no. 1, p. 3, 1937.

[SHI 05] SHI K. *et al.*, "Structure of peptide sex pheromone receptor PrgX and PrgX/pheromone complexes and regulation of conjugation in *Enterococcus faecalis*", *Proc. Natl. Acad. Sci. U.S.A.*, vol. 102, no. 51, pp. 18596–18601, 2005.

[SID 79] SIDEN I., DALHAMMAR G., TELANDER B. *et al.*, "Virulence factors in *Bacillus thuringiensis*: purification and properties of a protein inhibitor of immunity in insects", *Microbiology*, vol. 114, no. 1, pp. 45–52, 1979.

[SIF 02] SIFRI C.D. *et al.*, "Virulence effect of *Enterococcus faecalis* protease genes and the quorum-sensing locus fsr in *Caenorhabditis elegans* and mice", *Infect. Immun.*, vol. 70, no. 10, pp. 5647–5650, 2002.

[SIN 71] SINENSKY M., "Temperature control of phospholipid biosynthesis in *Escherichia coli*", *J. Bacteriol.*, vol. 106, no. 2, pp. 449–455, 1971.

[SLA 02] SLAMTI L., LERECLUS D., "A cell–cell signaling peptide activates the PlcR virulence regulon in bacteria of the *Bacillus cereus* group", *EMBO J.*, vol. 21, no. 17, pp. 4550–4559, 2002.

[SLA 14] SLAMTI L., PERCHAT S., HUILLET E. *et al.*, "Quorum sensing in *Bacillus thuringiensis* is required for completion of a full infectious cycle in the insect", *Toxins*, vol. 6, no. 8, pp. 2239–2255, 2014.

[SMI 04] SMITH J.L., FRATAMICO P.M., NOVAK J.S., "Quorum sensing: a primer for food microbiologists", *J. Food Prot.*, vol. 67, no. 5, pp. 1053–1070, 2004.

[SMI 06] SMITH D. *et al.*, "Variations on a theme: diverse N-acyl homoserine lactone-mediated quorum sensing mechanisms in Gram-negative bacteria", *Sci. Prog.*, vol. 89, nos 3–4, pp. 167–211, 2006.

[SMI 09] SMIT B.A., ENGELS W.J., SMIT G., "Branched chain aldehydes: production and breakdown pathways and relevance for flavour in foods", *Appl. Microbiol. Biotechnol.*, vol. 81, no. 6, pp. 987–999, 2009.

[SMI 14] SMID E., KLEEREBEZEM M., "Production of aroma compounds in lactic fermentations", *Annu. Rev. Food Sci. Technol.*, vol. 5, pp. 313–326, 2014.

[SOM 69] SOMKUTI G., BABEL F., "Hydrolytic breakdown of casein by a proteinase of *Streptococcus faecalis* var. liquefaciens", *J. Dairy Sci.*, vol. 52, no. 8, pp. 1186–1191, 1969.

[SPA 10] SPANO G. *et al.*, "Biogenic amines in fermented foods", *Eur. J. Clin. Nutr.*, vol. 64, pp. S95–S100, 2010.

[SPE 03] SPERANDIO V., TORRES A.G., JARVIS B. *et al.*, "Bacteria–host communication: the language of hormones", *Proc. Natl. Acad. Sci.*, vol. 100, no. 15, pp. 8951–8956, 2003.

[STE 68] STEINHAUER J.E., Quantitative determination of certain short-chain acids in frozen whole eggs by gas–liquid chromatography, PhD thesis, Michigan State University Department of Food Science, 1968.

[STE 02] STENFORS L.P., MAYR R., SCHERER S. *et al.*, "Pathogenic potential of fifty *Bacillus weihenstephanensis* strains", *FEMS Microbiol. Lett.*, vol. 215, no. 1, pp. 47–51, 2002.

[STE 04] STEPANIAK L., "Dairy enzymology", *Int. J. Dairy Technol.*, vol. 57, nos 2–3, pp. 153–171, 2004.

[STE 09] STELIOS K., PARASKEVI S., THEOPHILOS M. *et al.*, "Study of organic acids, volatile fraction and caseins of a new Halloumi-type cheese during ripening in whey brine", *Int. J. Food Sci. Technol.*, vol. 44, no. 2, pp. 297–304, 2009.

[STI 78] STILES M., RAMJI N., NG L-K. *et al.*, "Incidence and relationship of group D streptococci with other indicator organisms in meats", *Can. J. Microbiol.*, vol. 24, no. 12, pp. 1502–1508, 1978.

[STI 87] STIJVE T., DISERENS J., "Simplified method for the routine determination of organic-acids in eggs and egg products", *Dtsch. Lebensm.-Rundsch.*, vol. 83, no. 2, pp. 44–47, 1987.

[STI 89] STILES M. *et al.*, "Less recognized or presumptive foodborne pathogenic bacteria", *Foodborne Bact. Pathog.*, pp. 673–733, 1989.

[STI 97] STILES M.E., HOLZAPFEL W.H., "Lactic acid bacteria of foods and their current taxonomy", *Int. J. Food Microbiol.*, vol. 36, no. 1, pp. 1–29, 1997.

[STO 02] STOODLEY P., SAUER K., DAVIES D.G. *et al.*, "Biofilms as complex differentiated communities", *Annu. Rev. Microbiol.*, vol. 56, no. 1, pp. 187–209, 2002.

[STU 02] STURME M.H., KLEEREBEZEM M., NAKAYAMA J. *et al.*, "Cell to cell communication by autoinducing peptides in Gram-positive bacteria", *Antonie Van Leeuwenhoek*, vol. 81, no. 1, pp. 233–243, 2002.

[SUÁ 13] SUÁREZ C., ESPARIZ M., BLANCATO V.S. *et al.*, "Expression of the agmatine deiminase pathway in *Enterococcus faecalis* is activated by the AguR regulator and repressed by CcpA and PTSMan systems", *PLoS One*, vol. 8, no. 10, p. e76170, 2013.

[SUM 07] SUMAN M., RIANI G., DALCANALE E., "MOS-based artificial olfactory system for the assessment of egg products freshness", *Sens. Actuators B Chem.*, vol. 125, no. 1, pp. 40–47, 2007.

[SUZ 00] SUZZI G. *et al.*, "A survey of the enterococci isolated from an artisanal Italian goat's cheese (semicotto caprino)", *J. Appl. Microbiol.*, vol. 89, no. 2, pp. 267–274, 2000.

[SUZ 03] SUZZI G., GARDINI F., "Biogenic amines in dry fermented sausages: a review", *Int. J. Food Microbiol.*, vol. 88, no. 1, pp. 41–54, 2003.

[SWI 01] SWIFT S., DOWNIE J.A., WHITEHEAD N.A. *et al.*, "Quorum sensing as a population-density-dependent determinant of bacterial physiology", *Adv. Microb. Physiol.*, vol. 45, pp. 199–270, 2001.

[TAV 98] TAVARIA F.K., MALCATA F.X., "Microbiological characterization of Serra da Estrela cheese throughout its Appellation d'Origine Protegee region", *J. Food Prot.*, vol. 61, no. 5, pp. 601–607, 1998.

[TAV 02] TAVARIA F., DAHL S., CARBALLO F. *et al.*, "Amino acid catabolism and generation of volatiles by lactic acid bacteria", *J. Dairy Sci.*, vol. 85, no. 10, pp. 2462–2470, 2002.

[TEC 13] TECHER M-C., BARON F., JAN S., *Microbial spoilage of eggs and egg products*, vol. 69 supplement, World's poultry science association - Italian branch, 2013.

[TEC 14] TECHER C. *et al.*, "Global overview of the risk linked to the *Bacillus cereus* group in the egg product industry: identification of food safety and food spoilage markers", *J. Appl. Microbiol.*, vol. 116, no. 5, pp. 1344–1358, 2014.

[TEC 15] TECHER C., DAOUD A., MADEC M.-N. *et al.*, "Microbial quality of industrial liquid egg white: assumptions on spoiling lssues in egg-based chilled desserts: bacteria from liquid egg white", *J. Food Sci.*, vol. 80, no. 2, pp. M389–M398, 2015.

[THA 90] THAM W., KARP G., DANIELSSON-THAM M-L., "Histamine formation by enterococci in goat cheese", *Int. J. Food Microbiol.*, vol. 11, nos 3–4, pp. 225–229, 1990.

[TIT 98] TITBALL R.W., "Bacterial phospholipases", *J. Appl. Microbiol.*, vol. 84, no. S1, 1998.

[TOD 05] TODOROV S.D., WACHSMAN M.B., KNOETZE H. *et al.*, "An antibacterial and antiviral peptide produced by *Enterococcus mundtii* ST4V isolated from soya beans", *Int. J. Antimicrob. Agents*, vol. 25, no. 6, pp. 508–513, 2005.

[TOD 09] TODOROV S.D., VON MOLLENDORFF J.W., MOELICH E. *et al.*, "Evaluation of potential probiotic properties of *Enterococcus mundtii*, its survival in boza and in situ bacteriocin production", *Food Technol. Biotechnol.*, vol. 47, no. 2, 2009.

[TOM 08] TOMÉ E., GIBBS P.A., TEIXEIRA P.C., "Growth control of Listeria innocua 2030c on vacuum-packaged cold-smoked salmon by lactic acid bacteria", *Int. J. Food Microbiol.*, vol. 121, no. 3, pp. 285–294, 2008.

[TRO 87] TROVATELLI L., SCHIESSER A., MASSA S., "Identification and significance of enterococci in hard cheese made from raw cow and sheep milk", *Milchwissenschaft*, vol. 42, no. 11, pp. 717–719, 1987.

[TSA 93] TSAKALIDOU E., MANOLOPOULOU E., TSILIBARI V. *et al.*, "Esterolytic activities of *Enterococcus durans* and *Enterococcus faecium* strains isolated from Greek cheese", *Neth. Milk Dairy J.*, vol. 47, pp. 145–145, 1993.

[TSA 94a] TSAKALIDOU E. *et al.*, "The combined use of whole-cell protein extracts for the identification (SDS-PAGE) and enzyme activity screening of lactic acid bacteria isolated from traditional Greek dairy products", *Syst. Appl. Microbiol.*, vol. 17, no. 3, pp. 444–458, 1994.

[TSA 94b] TSAKALIDOU E., DALEZIOS I., KALANTZOPOULOS G., "Isolation and partial characterization of an intracellular esterase from *Enterococcus faecium* ACA-DC 237", *J. Biotechnol.*, vol. 37, no. 3, pp. 201–208, 1994.

[TUR 83] TURNBULL P., KRAMER J.M., "Non-gastrointestinal *Bacillus cereus* infections: an analysis of exotoxin production by strains isolated over a two-year period", *J. Clin. Pathol.*, vol. 36, no. 10, pp. 1091–1096, 1983.

[URD 95] URDANETA D., RAFFE D., FERRER A. *et al.*, "Short-chain organic acids produced on glucose, lactose, and citrate media by *Enterococcus faecalis*, *Lactobacillus casei*, and *Enterobacter aerogenes* strains", *Bioresour. Technol.*, vol. 54, no. 2, pp. 99–103, 1995.

[VAL 10] VALENZUELA A.S., BENOMAR N., ABRIOUEL H. *et al.*, "Isolation and identification of *Enterococcus faecium* from seafoods: Antimicrobial resistance and production of bacteriocin-like substances", *Food Microbiol.*, vol. 27, no. 7, pp. 955–961, October 2010.

[VAN 71] VAN DER WERF P., ORLOWSKI M., MEISTER A., "Enzymatic conversion of 5-oxo-L-proline (L-pyrrolidone carboxylate) to L-glutamate coupled with cleavage of adenosine triphosphate to adenosine diphosphate, a reaction in the γ-glutamyl cycle", *Proc. Natl. Acad. Sci.*, vol. 68, no. 12, pp. 2982–2985, 1971.

[VAN 06] VAN DEN BERGHE E., DE WINTER T., DE VUYST L. *et al.*, "Enterocin A production by *Enterococcus faecium* FAIR-E 406 is characterised by a temperature-and pH-dependent switch-off mechanism when growth is limited due to nutrient depletion", *Int. J. Food Microbiol.*, vol. 107, no. 2, pp. 159–170, 2006.

[VAN 08] VAN DER VOORT M., KUIPERS O.P., BUIST G. *et al.*, "Assessment of CcpA-mediated catabolite control of gene expression in *Bacillus cereus* ATCC 14579", *BMC Microbiol.*, vol. 8, no. 1, p. 62, 2008.

[VAS 08] VASILOPOULOS C. *et al.*, "Evaluation of the spoilage lactic acid bacteria in modified-atmosphere-packaged artisan-type cooked ham using culture-dependent and culture-independent approaches", *J. Appl. Microbiol.*, vol. 104, no. 5, pp. 1341–1353, 2008.

[VAS 16] VASIEE A., BEHBAHANI B.A., YAZDI F.T. *et al.*, "Optimization of the production conditions of the lipase produced by *Bacillus cereus* from rice flour through Plackett-Burman Design (PBD) and response surface methodology (RSM)", *Microb. Pathog.*, vol. 101, pp. 36–43, 2016.

[VER 97] VERGER R., "Interfacial activation'of lipases: facts and artifacts", *Trends Biotechnol.*, vol. 15, no. 1, pp. 32–38, 1997.

[VIL 94] VILLANI F., COPPOLA S., "Selection of enterococcal strains for water-buffalo Mozzarella cheese manufacture", *Ann. Microbiol. Ed Enzimologia Italy*, 1994.

[VIT 16] VITHANAGE N.R., DISSANAYAKE M., BOLGE G. *et al.*, "Biodiversity of culturable psychrotrophic microbiota in raw milk attributable to refrigeration conditions, seasonality and their spoilage potential", *Int. Dairy J.*, vol. 57, pp. 80–90, June 2016.

[VON 98] VON STETTEN F., FRANCIS K.P., LECHNER S. *et al.*, "Rapid discrimination of psychrotolerant and mesophilic strains of the *Bacillus cereus* group by PCR targeting of 16S rDNA", *J. Microbiol. Methods*, vol. 34, no. 2, pp. 99–106, 1998.

[VON 99] VON STETTEN F., MAYR R., SCHERER S., "Climatic influence on mesophilic *Bacillus cereus* and psychrotolerant *Bacillus weihenstephanensis* populations in tropical, temperate and alpine soil", *Environ. Microbiol.*, vol. 1, no. 6, pp. 503–515, 1999.

[VRI 05a] VRIES Y.P., ATMADJA R.D., HORNSTRA L.M. *et al.*, "Influence of glutamate on growth, sporulation, and spore properties of *Bacillus cereus* ATCC 14579 in defined medium", *Appl. Environ. Microbiol.*, vol. 71, no. 6, pp. 3248–3254, 2005.

[VRI 05b] VRIES Y.P., HORNSTRA L.M., ATMADJA R.D. *et al.*, "Deletion of *sigB* in *Bacillus cereus* affects spore properties", *FEMS Microbiol. Lett.*, vol. 252, no. 1, pp. 169–173, 2005.

[WAL 70] WALLACE D., HARMON L., "Intracellular protease from *Streptococcus durans*", *J. Dairy Sci.*, vol. 53, no. 4, pp. 394–402, 1970.

[WAR 83] WARD O.P., "Proteinases", in FORGARTY W.H. (ed.), *Appl. Sci. Publishers, New York*, Applied Science Publication, New York, pp. 251–317, 1983.

[WES 90] WESSELS D., JOOSTE P., MOSTERT J., "Technologically important characteristics of *Enterococcus* isolates from milk and dairy products", *Int. J. Food Microbiol.*, vol. 10, nos 3–4, pp. 349–352, 1990.

[WHI 00] WHITFIELD F.B., JENSEN N., SHAW K.J., "Role of *Yersinia intermedia* and *Pseudomonas putida* in the development of a fruity off-flavour in pasteurized milk", *J. Dairy Res.*, vol. 67, no. 4, pp. 561–569, 2000.

[WIL 07a] WILLIAMS P., "Quorum sensing, communication and cross-kingdom signalling in the bacterial world", *Microbiology*, vol. 153, no. 12, pp. 3923–3938, 2007.

[WIL 07b] WILLIAMS P., WINZER K., CHAN W.C. *et al.*, "Look who's talking: communication and quorum sensing in the bacterial world", *Philos. Trans. R. Soc. Lond. B Biol. Sci.*, vol. 362, no. 1483, pp. 1119–1134, 2007.

[WIN 03] WINZER K., HARDIE K.R, WILLIAMS P., "LuxS and autoinducer-2: their contribution to quorum", *Adv Appl Microbiol*, vol. 53, p. 291, 2003.

[WOO 88] WOOD S.L., WAITES W., "Factors affecting the occurrence of *Bacillus cereus* in liquid whole egg", *Food Microbiol.*, vol. 5, no. 2, pp. 103–107, 1988.

[WUN 14] WUNDERLICHOVÁ L., BUŇKOVÁ L., KOUTNỲ M. *et al.*, "Formation, degradation, and detoxification of putrescine by foodborne bacteria: a review", *Compr. Rev. Food Sci. Food Saf.*, vol. 13, no. 5, pp. 1012–1030, 2014.

[YAM 10] YAMAKAWA Y., NAU F., "Valeur nutritionnelle et allergénicité", *Science et technologie de l'œuf Vol. 2. De l'œuf aux ovoproduits*, Lavoisier, Paris, p. 177, 2010.

[YAN 85] YAN L., LANGLOIS B., O'LEARY J. *et al.*, "Purification and characterization of four extracellular proteases isolated from raw milk psychrotrophs", *J. Dairy Sci.*, vol. 68, no. 6, pp. 1323–1336, 1985.

[YAN 98] YANO Y., NAKAYAMA A., ISHIHARA K. *et al.*, "Adaptive changes in membrane lipids of barophilic bacteria in response to changes in growth pressure", *Appl. Environ. Microbiol.*, vol. 64, no. 2, pp. 479–485, 1998.

[YOO 08] YOON M.Y., KIM Y.J., HWANG H-J., "Properties and safety aspects of *Enterococcus faecium* strains isolated from Chungkukjang, a fermented soy product", *LWT – Food Sci. Technol.*, vol. 41, no. 5, pp. 925–933, 2008.

[YOR 71] YORK L.R., "Microbial counts and organic acid quantitation as quality indices of egg products", Michigan State University, 1971.

[YOR 72] YORK L., DAWSON L., "Organic acid accumulation in egg products inoculated with known bacterial cultures", *Poult. Sci.*, vol. 51, no. 4, pp. 1244–1247, 1972.

[YOU 05] YOUSIF N.M.K. *et al.*, "Molecular characterization, technological properties and safety aspects of enterococci from "Hussuwa", an African fermented sorghum product", *J. Appl. Microbiol.*, vol. 98, no. 1, pp. 216–228, 2005.

[ZEN 05a] ZENDO T. *et al.*, "Identification and production of a bacteriocin from *Enterococcus mundtii* QU 2 isolated from soybean", *J. Appl. Microbiol.*, vol. 99, no. 5, pp. 1181–1190, 2005.

[ZEN 05b] ZENEWICZ L.A., WEI Z., GOLDFINE H. *et al.*, "Phosphatidylinositol-specific phospholipase C of *Bacillus anthracis* down-modulates the immune response", *J. Immunol.*, vol. 174, no. 12, pp. 8011–8016, 2005.

[ZIG 06] ZIGHA A., ROSENFELD E., SCHMITT P. *et al.*, "Anaerobic cells of *Bacillus cereus* F4430/73 respond to low oxidoreduction potential by metabolic readjustments and activation of enterotoxin expression", *Arch. Microbiol.*, vol. 185, no. 3, pp. 222–233, 2006.

[ZIG 07] ZIGHA A., ROSENFELD E., SCHMITT P. *et al.*, "The redox regulator Fnr is required for fermentative growth and enterotoxin synthesis in *Bacillus cereus* F4430/73", *J. Bacteriol.*, vol. 189, no. 7, pp. 2813–2824, 2007.

3

Characterizing the Spoilage of Egg Products using Targeted and Non-targeted Approaches

3.1. Introduction

An egg product is a complex matrix, mainly composed of water, proteins and lipids. It also contains carbohydrates, vitamins, minerals and many other compounds in varying concentrations. This chemical complexity makes it difficult to analyze, and most of the protocols are mainly concerned with the albumen and vitellus fractions, aside from the whole egg product. The hygienic quality and freshness of eggs can be assessed according to physico-chemical or microbiological criteria, imposed by the regulations in force. The state provides procedures and recommendations for the production and marketing of such products. Criteria and procedures vary from region to region, for example European legislation forbids the breaking of eggs by centrifugation for eggs intended for the food industry whereas American

---

Chapter written by Rémy COAT, Boris GOUILLEUX, Gérald THOUAND, Ali ASSAF, Abdellah ARHALIASS, Jack LEGRAND, Anne THIERRY, Estelle MARTINEAU, Frédérique COURANT, Patrick GIRAUDEAU and Olivier GONÇALVES.

legislation allows the possibility of breaking eggs by centrifugation if they are clean, disinfected and intact and if they ultimately undergo heat treatment (baking etc.). In addition to the procedures and recommendations concerning the production process, the hygienic quality of egg products is ensured by controlling the physico-chemical and microbiological quality of the raw materials used and the finished products. European legislation thus provides maximum levels for certain compounds such as lactic acid or succinic acid at different points in the production chain.

Besides these legislative aspects of the specifications required in order to be placed onto the market, the spoilage of egg products during their manufacturing processes and their storage is a field of study in itself. Even if different potential markers of the hygienic quality or freshness of egg products have been proposed in recent decades, neither the latter nor the analytical specifications derived from the legislation are used routinely for detection or research. Spoilage of liquid whole egg products (LEs) is, therefore, not currently characterized by specific analyses. The deterioration is thus observed in production where it sometimes causes a degradation of the organoleptic properties of the products (color, odor, texture, etc.) as well as modifications to their physico-chemical characteristics (pH, rheology, etc.). The analytical approaches derived from the regulatory control standards and some more recently developed for the characterization of the quality of liquid whole eggs are sometimes used for the study of spoilage, but too few approaches have been developed specifically for this purpose. Moreover, these techniques are often long and expensive, sometimes requiring specific equipment and are very limited since they are designated for a given target.

Some rapid approaches based on the latest technical advances in analytical chemistry could partially replace conventional methods of routine control of the quality of LEs, and especially complement them in the context of the study of spoilage. These highly resolutive "omic" approaches would more accurately reflect the state of the egg matrix, since they are capable of analyzing many thousands of compounds at once. This is particularly the case of approaches resulting from metabolomics. The term "metabolomics" was introduced in 1999 by J. Nicoholson in the United Kingdom, and then by O. Fiehnin in 2001 as "quantitative measurement of the dynamic multiparametric metabolic response of living systems to pathophysiological stimuli or genetic modification" [ZHA 14]. Like transcriptomics and proteomics, metabolomics is a post-genomic discipline.

The goal of metabolomics is the study of the metabolome. This represents all of the metabolites present in a biological system: organism, cell, tissue and biological fluid (plasma, urine). The metabolites are organic molecules with a molecular weight of less than 1,000 Da such as amino acids, organic acids, monosaccharides and small peptides. Nucleic acids and proteins are not metabolites due to their excessively high molecular weight. Since metabolites are the end products of cellular enzymatic machinery, they can be considered as the ultimate response to genetic and/or environmental variations of a biological system [FIE 02]. However, analyzing the metabolome is not easy to implement since we must try to take into account all of its complexity. The number of metabolites listed in living organisms is estimated at over 50,000.

Different types of metabolomic approaches may be suggested for characterizing the metabolism of an organism [FIE 02]:

– targeted metabolomics (~ 10 metabolites)

– metabolic profiling (~100 metabolites)

– metabolomics (~1000 metabolites)

– metabolic fingerprint (>1000 metabolites).

The targeted analysis of metabolites can be defined as a detailed analysis, focused on one or a small number of metabolites of a particular metabolic pathway, which may require fragmentation of the sample to avoid interferences induced by other compounds involved. It is a quantitative approach.

Metabolic profiling is defined in the same way as previously, consisting of identifying and quantifying a small number of metabolites, except in this case, unlike targeted metabolomics, the compounds may belong to different metabolic pathways, with the latter being more numerous. The sample preparation focuses on the chemical properties of the desired compounds in order to limit matrix effects.

Metabolomics is the identification and quantification of the largest number of possible compounds in a sample, and the identification methods used are usually automated. The information obtained is more comprehensive but less accurate than in profiling or targeted analysis.

In contrast to these three approaches whose goal is the precise identification and quantification of a variable number of metabolites, the objective of the metabolic fingerprint is to compare metabolic profiles from two experimental conditions in order to establish signatures specific to the phenomenon studied [KOP 04]. This is the non-targeted study of a large number of metabolites, and these metabolites are therefore not quantified in an absolute way; this is called relative quantification.

The advantage of so-called "targeted" metabolomics when compared to the "non-targeted" metabolomics is the simplified obtainment of a precise quantification; its main disadvantage is its "limitation in terms of the extent of the analysis" [YI 16].

In order to be able to apply a metabolic profiling approach to the study of LE spoilage, it is necessary to control, in addition to the culture conditions of the organism or organisms used to reproduce the microbiological spoilage, all of the sample preparation steps: that is, the EL preparation and sampling methods, the pre-treatment of the samples and the extraction of the metabolites according to the techniques used (deproteinization, drying, fractionation, addition of an internal standard etc.). Choosing the analytical technique is also important since it must make it possible to estimate the diversity of the metabolic footprint. Analyzing the raw data and the results are also delicate steps since the results obtained are clearly influenced by the techniques or the quality of the databases used. All of these key operations need to be mastered to avoid the accumulation of bias during data processing operations.

In metabolomics, the most commonly used techniques are liquid chromatography coupled with mass spectrometry (LC-MS) and nuclear magnetic resonance (NMR) because of their high analytical resolution. However, a less resolutive analysis by gas chromatography coupled with mass spectrometry GC-MS can be an option (Figure 3.1). Although this technique has a lower performance, it is nevertheless robust and easy to access, which allows the quantification of most of the compounds and thus a better understanding of the phenomena involved in the redirection of carbon fluxes.

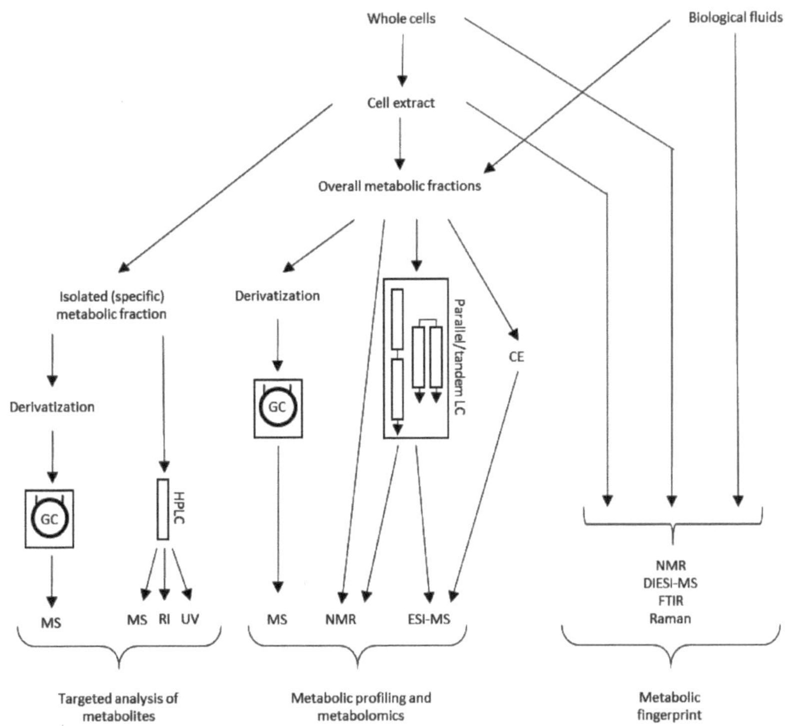

Figure 3.1. *Metabolomics techniques: technologies used for methods and sampling (inspired by the work of [DUN 05])*

In this chapter, we will discuss analysis methods dealing specifically with the liquid whole egg product (mixture of unchanged proportions of yolk and egg white) when these are referenced. First, the current criteria used for the characterization of LEs will be presented, while emphasizing their analytical limitations. Then, we will discuss the contribution of new approaches resulting from vibrational spectroscopy and nuclear magnetic resonance or mass spectrometry for the physicochemical characterization of the spoilage of the LE matrix. The potential of using these new approaches will also be evaluated specifically for the analysis of LE spoilage.

3.2. Regulatory analyses for measuring the quality of the liquid whole egg are not sufficient to estimate their spoilage

Different methods of invasive and non-invasive analyses exist and are used to evaluate the age and freshness of shell eggs (eggs that have not been separated from their shells), and include the measurement of the air space that allows the classification of eggs by category in France (A or B) and the Haugh unit, used to obtain the red label in France and as a freshness indicator in US legislation. However, once separated from its shell and mixed, the egg product can no longer be subjected to these simple tests and other health and safety criteria are applied.

The production centers of French egg products are subject to French and European legislation; the production circuits are controlled and subject to approval depending on the type of production (liquid egg, processed egg product, etc.). The production facilities are designed to limit the risk of contamination (cleaning, disinfection, separation of the operations listed in regulation (EC) no. 853/2004, Appendix III, section X, Chapter II, point I). Production operations such as pasteurization are controlled by the producer, but the latter is subject to criteria with regard to the hygienic quality of the finished products (for pasteurization for example, at least seven log10 reductions in bacteria of the genus *Salmonella* are expected; in other words, the number of bacteria must be 10 million times lower at the end of this step) [MIN 17]. The commercialization of egg products is also preceded by checks to inspect the state of freshness of the products, both physico-chemically and microbiologically.

The regulatory criteria listed below are extracted from texts referenced in the report on Vademecum sectorial production of egg products and liquid whole eggs (Version 2.0, February 2017) ([MIN 17]), available on the French Ministry of Agriculture, Agribusiness and Forestry website. These are mainly texts resulting from EC regulations no. 2073/2005 of November 15, 2005 on the microbiological criteria applicable to foodstuff [EUR 05] and (EC) no. 853/2004 of the European Parliament and Council, April 29, 2004 laying down specific hygiene rules for food of animal origin [EUR 04], and the UNECE standard egg-2 concerning the marketing and commercial quality control of egg products – 2010 edition [UNE 10].

3.2.1. *Physico-chemical criteria*

Some specifications are provided by legislation to ensure the quality of raw materials used as well as marketed products. Regulation (EC) no. 853/2004 [EUR 04] provides analytical specifications that egg products must meet for the European market:

– the concentration of 3-hydroxy-butyric acid must not exceed 10 mg/kg per dry matter of unmodified egg product;

– the lactic acid content of the raw material used to make the egg products must not exceed 1 g/kg of dry matter;

– the amount of eggshell residues, egg membranes and other possible particles in the processed egg product must not exceed 100 mg/kg of egg product.

The UNECE standard egg-2 [UNE 10] specifies that egg products obtained by conventional techniques and without the use of special processes to modify the properties and/or composition of the product must satisfy a certain number of requirements, listed in Table 3.1, which are partly redundant with those of regulation (EC) no. 853/2004 [EUR 04].

	Product						
	Whole egg		Egg yolk		Egg white		
	Liquid and frozen	Dried	Liquid and frozen	Dried	Liquid and frozen	Dried	
						Tray dried	Spray dried
Minimum solid element content (% dry weight)	22	95	40	95	10.5	84	92
Minimum fat content (% dry weight)	9.8	39	25	55	0.05	0.05	
Minimum protein content (% dry weight)	10.5	45	15	33	10	71	75
Foreign bodies	No particle greater than 1 mm in 100 g and maximum of 100 mg*kg						
Minimal concentration of hydrogen ions (pH)	7	7.5	5.9	6	8.5	4	

Maximum quantity of beta-hydroxybutyric acid(1) (mg/kg)	10	10	10	10	10	10
Maximum quantity of lactic acid (mg/kg)	1000	-	1000	-	1000	-
Maximum quantity of succinic acid(2) (mg/kg)	25	-	25	-	25	-

Note: In the case of dried egg products, the weight ratio of fat and egg white as solids is calculated.
(1) Lactic acid (used only for processing), succinic acid and beta-hydroxybutyric acid are considered as solid elements. The maximum weight of shells, egg membranes and other particles present in egg products is 100 mg/kg of egg product.
(It may be necessary to express the solids content as dry matter content. Solids could be interpreted as the residue of liquid whole egg filtration. When the liquid whole eggs are simply dehydrated, the dry residue will be dry matter. Since liquid whole eggs do not contain any suspended solids, these liquid eggs that have been dried should be indicated as dry matter.)
(2) The succinic acid limit must be indicated in order to prevent rejects from hatcheries being used in food-grade egg products for the food industry; this is a quality parameter.

Table 3.1. *Physical and chemical indicators of typical egg products*

3.2.2. *Microbiological criteria*

Regulation (EC) no. 853/2004 [EUR 04] clearly indicates that the food business operators must ensure that all operations are carried out in a manner that avoids contamination during the production, handling and storage of egg products and provides a number of guidelines and protocols to ensure the hygienic quality of the products.

Besides these precautions and the pasteurization step, liquid whole egg products are subjected to microbiological analyses on each batch of product in order to meet certain specifications, including the absence of bacteria of the genus *Salmonella* in 25 g of product (Table 3.2) and the presence of *Enterobacteriaceae* lower than a defined threshold (Table 3.3).

– Absence of the bacterial genus *Salmonella* in 25 g of product.

Food category	Micro-organisms/ toxins, metabolites	Sampling plans (1)		Limits		Reference analysis method (2)	Application stage
		n	c	m	M		
1.14: Egg products, except products whose manufacturing processor composition prevents risk of salmonella	*Salmonella*	5	0	Absence in 25 g		EN/ISO 6579	Marketed products during their shelf life

(1) n = number of units in the sample; c = number of units in the sample giving values between m and M.

(2) It is necessary to use the most recent edition.

Table 3.2. *Food safety criteria (regulation (EC) no. 2073/2005 – Appendix III, Chapter 1.14) [EUR 05]*

Food category	Micro-organisms	Sampling plans(1)		Limits (ufc/g or ml)		Reference analysis method(2)	Application stage	Action in case of unsatisfactory results
		n	c	m	M			
2.3 Egg products	*Entero-bacteriaceae*	5	2	10	100	ISO 215 28-2	End of manufacturing process	Effective control of heat treatment and prevention of recontamination

(1) n = number of units in sample; c = number of units in sample giving results between m and M.

(2) It is necessary to use the most recent edition.

The limits indicated apply to each sample unit analyzed.

The analysis results reveal the microbiological quality of the controlled process.

Enterobacteria in egg products:

– satisfactory quality when all values observed are ≤ m;

– acceptible quality when a maximum of c/n values are between m and M, and the rest of the values observed are ≤ m;

– unsatisfactory quality when one or several values observed are > M or when more than c/n values are between m and M.

Table 3.3. *Hygiene criteria of processes – Egg products (regulation (EC) no. 2073/2005 – Appendix III, Chapter 2.3) [EUR 05]*

3.2.3. *Limitations of current analyses and the importance of finding complementary approaches*

The criteria listed in sections 2.1 and 2.2 provide protection against the use of raw materials that are highly contaminated or have a specific risk factor (eggs that have been rejected from incubators). They also provide protection against certain specific germs that pose risks to public health (*Salmonella*). However, current analyses do not ensure the freshness of a product, the optimization of its optimal shelf life or the detection of the microbiological spoilage risks of the products during their storage. This type of spoilage leads to frequent loss of product. The responsibility of the manufacturer is often engaged by the legislation which imposes obligations for results rather than for means, and the producers have a non-contractual obligation to carry out a certain number of additional analyses, on top of the regulatory controls, including the total flora count on PCA medium and the search for bacteria of the genus *Listeria* or even *Bacillus* [NAU 10].

These microbiological analyses have been supplemented in recent years by physic-chemical approaches allowing the identification of degradation markers of egg products during their manufacturing and storage process. Some of these markers are not used frequently because of their thermolability or their natural variability in the raw material (uridine, pyroglutamic acid). However, some of them, such as furosine, have interesting potential, [ROS 13]. Currently, there are fewer analytical methods that can be used for the analysis of freshness and the detection of spoilage of liquid egg products than there are for shell eggs, since some of these techniques are not appropriate on a broken and mixed egg.

Table 3.4 lists the identified or potential degradation markers for egg products in the literature. The shaded boxes correspond to the markers whose analysis is now included in the legislative criteria relating to the production, processing or commercialization of egg products.

Markers of bacterial degradation	**Organic acids**
Lactic acid [LIT 82, HID 08, SAL 04]	Lactic acid [LIT 82, HID 08, SAL 04]
Succinic acid [LIT 82, HID 08, SAL 04]	Succinic acid [LIT 82, HID 08, SAL 04]
Acetic acid [LIT 82, HID 08, SAL 04]	Hydroxy-butyric acid [LIT 82, SAL 72, HEA 76, PAR 80, ROB 75]
Uracil [HID 08, MOR 89, UTA 85]	Acetic acid [LIT 82, HID 08, SAL 04]
Glucose [WAR 15]	**Pyrimidine bases**
Autoinducer-2 [RUL 15]	Uracil [HID 08, MOR 89, UTA 85]
AHL [CHU 11]	**Animo acids**
Bacterial fatty acids [OLE 62, LEC 77, LAN 11]	Pyroglutamic acid [ROS 95, LUC 96]
Ethanol [WAR 15]	Lysine [ERB 07, HID 06, HID 95]
Hydroxy-butyric acid [LIT 82, SAL 72, HEA 76, PAR 80, ROB 75]	**Compounds of Maillard reactions**
Markers of physico-chemical degradation	Furosine [ERB 07, HID 06, HID 95]
Pyroglutamic +acid [ROS 95, LUC 96]	**Reducing sugars**
Lysine [ERB 07, HID 06, 95]	Glucose [WAR 15, CAB 05, FRI 96]
Furosine [ERB 07, HID 06, 95]	**Quorum Sensing Molecules**
Glucose [CAB 05, FRI 96]	*Autoinducer*-2 [RUL 15]
Cholesterol and COPs [CAB 05, ZAR 14, GUA 94]	*AHL* [CHU 11]
Tocopherol [CAB 05, GAL 01, WAH 93]	**Lipids and oxidized lipids**
Retinol [CAB 05, GAL 01, WAH 93]	Cholesterol and COPs [CAB 05, ZAR 14, GUA 94]
	Vitamins
	Tocopherol [CAB 05, GAL 01, WAH 93]
	Retinol [CAB 05, GAL 01, WAH 93]
	Other compounds
	Bacterial fatty acids [OLE 62, LEC 77, LAN 11]
	Ethanol [WAR 15]

Table 3.4. *Spoilage markers for potential (in italics) and confirmed (in normal font) LE products identified in the bibliography. The markers are categorized as biotic or abiotic in the left colum and by chemical family in the column on the right*

Regulatory analyses ensure food safety and the quality of the products used but do not detect and/or avoid the spoilage of LEs. Many potential markers of degradation identified in the literature could now better assess the quality of these products as well as their potential stage of spoilage, thus avoiding losses of products related to microbiological contamination.

3.3. Vibrational spectroscopy allows the chemotyping of the main classes of the biomolecules present in the LE matrix

3.3.1. *The benefits of vibrational techniques*

These are non-invasive approaches that involve little or no sample preparation and can therefore be applied to an entire matrix. It is also possible to apply them on extracts of contrasting polarity. This advantage is, however, counterbalanced by the resolution or limited sensitivity of the chemical and biochemical information obtained.

3.3.2. *Infrared spectroscopy (IR)*

3.3.2.1. *General introduction to the principle of IR spectroscopy*

3.3.2.1.1. Infrared radiation

The principle of spectroscopy is based on the interaction between electromagnetic radiation and a sample, which results in energy modifications of the two contributors of the phenomenon. Infrared radiation was discovered in 1800 by William Herschel. This component of electromagnetic radiation can be described as the combination of a magnetic field and an electric field. They oscillate perpendicularly in phase according to a bisinusoidal in the direction of wave propagation. Infrared radiation can also be described as being associated with a massless flow of particles: *photons*.

These particles carry a certain amount of quantum energy whose value is given by the *Bohr* equation:

$$E = h\upsilon = h\frac{c}{\lambda} = hc\,\overline{\upsilon}$$

Where h = 6.62.10^{-34} J.s (Planck constant) and $\overline{\upsilon} = \frac{1}{\lambda}$ (wave number in cm^{-1}). The spectral ranges used in infrared spectrometry and which make it possible to identify the chemical bonds corresponding to the functional groups of different molecules, are those of the mid and the near infrared (Tables 3.5 and 3.6, respectively). Far infrared, corresponding to the last spectral range of IR, will not be discussed in this book.

ν	2.5 μm to 25 μm
Energy E	$7.95\cdot10^{-20}$ J (i.e. 0.4965 eV) to $7.95\cdot10^{-21}$ J (i.e. 0.04965 eV)
Frequency ν	$1.2\cdot10^{14}$ Hz to $1.2\cdot10^{13}$ Hz
Wave number $\bar{\upsilon}$	4000 cm^{-1} to 400 cm^{-1}

Table 3.5. *Mid-infrared spectral range*

ν	0.78μm to 2.5 μm
Energy E	$1.60.10^{-19}$ J (i.e. 1.5895 eV) to $7.95.10^{-20}$ J (i.e. 0.4965 eV)
Frequency ν	$3.8\ .10^{11}$ Hz to $1.2\ .10^{14}$ Hz
Wave number $\bar{\upsilon}$	12820 cm^{-1} to 4000 cm^{-1}

Table 3.6. *Near-infrared spectral range*

3.3.2.1.2. Infrared spectrometry is a molecular vibration spectroscopy

The total energy of molecules can be represented as the sum of three components: $E_{tot} = E_{elec} + E_{vibr} + E_{rot}$

E_{elec}: molecular electron energy associated with the orbitals of the peripheral electrons of the molecule (electrons involved in the formation of covalent bonds).

E_{vibr}: energy related to the vibrations of all of the atoms and bonds of the molecule.

E_{rot}: the rotational energy of the molecule around its center of gravity.

We are therefore interested in the transitions that take place when the molecule absorbs electromagnetic radiation in the infrared range, and in the fact that the energy of the latter is equal to the energy difference between two vibrational levels. The chemical bonds of the molecule vibrate around their equilibrium position. If the atoms of the molecule have different electronegativities, the molecule will form a permanent electric dipole. The infrared absorption will therefore correspond to a coupling between the electrical component of the electromagnetic wave and such a dipole.

3.3.2.1.3. Modeling vibrational interactions – harmonic oscillator model

The chemical bond connecting two atoms can be described as a stiffness spring k joining two bodies C_1 and C_2 of respective masses m_1 and m_2. If the two masses move a distance from the equilibrium position $x = \Delta x_1 + \Delta x_2$, the system oscillates around this position at a certain frequency which depends on the stiffness of the spring and the masses of the two bodies (Figure 3.2).

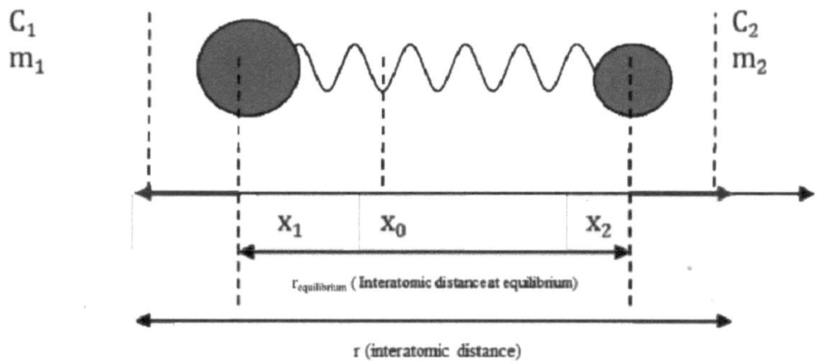

Figure 3.2. *Representation of a diatomic molecule in the form of a harmonic oscillator. Natural frequency of the oscillator* $\upsilon_{vibr} = \frac{1}{2\pi}\sqrt{\frac{k}{\mu}}$ *(k is the bond stiffness constant (N.m*$^{-1}$*); μ is the reduced mass of two atoms (kg)). Reduced mass of system:* $\mu = \frac{m_1 \times m_2}{m_1 + m_2}$. *Potential energy of oscillator:* $Ep = \frac{1}{2}kx^2$ *(with* $x = r - r_{equilibrium}$*). The potential energy describes a parabole whose minimum corresponds to the length of the spring at rest. For a color version of this figure, see www.iste.co.uk/goncalves/ovoproducts.zip*

3.3.2.1.4. Real situation: Morse curve

A chemical bond cannot be stretched to infinity because beyond a certain elongation; the bond is broken and the potential energy becomes zero when the atoms are distant. When the interatomic distance is reduced, forces of attraction come into play and they increase the stability of the system and decrease the potential energy. Finally, when the interatomic distance is reduced further, forces of repulsion come into play and reduce the stability of the system while increasing its potential energy. The Morse curve describes the variation of potential energy as a function of the internuclear distance (Figure 3.3).

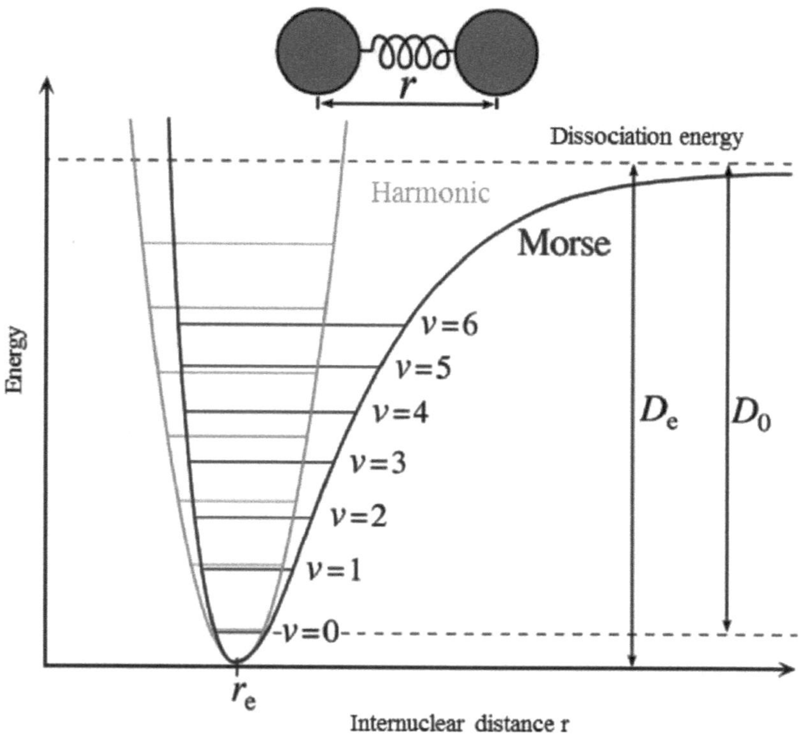

Figure 3.3. *Potential energy versus internuclear distance. r: harmonic oscillator model and Morse curve. De: dissociation energy of the bond. D0: actual dissociation energy of the bond. For a color version of this figure, see www.iste.co.uk/goncalves/ovoproducts.zip*

The lowest energy level that the molecule can reach is v = 0. The energy for this point is minimal; it is called "zero point energy" (even for absolute zero the molecule continues to vibrate). The dissociation energy D0 of the bond can be obtained by subtracting the zero point energy (E0) from the well depth. This Morse Function is $E(r)=k(1-e^{\beta(r-r_0)^2})$ with k (stiffness constant) = De. Regarding r_e, we can go back to the model of the harmonic oscillator and write: $E(r)=\frac{1}{2}De(r-r_e)^2$.

3.3.2.1.5. Quantum approach

The vibrational energy of the molecule can be described by the Schrödinger equation:

$$E = h\upsilon.\left(v + \frac{1}{2}\right) = \frac{h}{2\pi}\sqrt{\frac{k}{\mu}}\left(v + \frac{1}{2}\right)$$

v = quantum number of vibration that can take values of 0, 1, 2, etc.

υ = natural frequency of the oscillator

A transition will be permitted only if $\Delta v = \pm 1$, that is, if it occurs between two consecutive vibrational levels.

3.3.2.1.6. Case of a polyatomic molecule

The vibration of a polyatomic molecule composed of N atoms is relatively complex and can be decomposed into several simple modes of vibration. Locating an atom in space requires three coordinates: a molecule of N atoms therefore has 3N degrees of freedom. The degrees of freedom can be broken down into three modes: *translational mode, rotational mode* and *vibrational mode*. In general, a molecule has three degrees of translation and three degrees of rotation, so it has 3N-6 degrees of vibration. An absorption band will be visible on an infrared spectrum only if it causes a variation of the dipole moment of the molecule. We distinguish different modes of molecular vibration: the elongation vibrations and the deformation vibrations that can take place inside of the plane δ or outside of the plane γ.

3.3.2.1.7. The influence of the heterogeneity of the deposit

When an infrared spectrum is obtained for a heterogeneous product (for example containing particles or living cells that can behave like dielectric spheres), large spectral variations can be observed. Relevant work by

[ROM 06] clearly emphasizes the importance of these effects, which were largely ignored in many studies prior to this research. Indeed, the *scattering* effect of radiation produces large fluctuations superimposed on the spectrum [MIE 08]. The dielectric spheres are known to disperse electromagnetic radiation if the wavelength of light considered is comparable to the size of the dielectric sphere. This causes a sinusoidal shape distortion impacting the overall shape of the baseline of the infrared spectrum. Depending on the size and shape of the particles in the analyzed matrices, varying effects will be recorded.

3.3.2.1.8. Statistical processing of infrared spectra

There are different possible approaches to processing spectral data. In the case of simple matrices, it is possible, for example, to quantify classes of compounds with vibrators of interest using approaches such as multiple linear regression [LIC 02]. Food matrices, however, are often too complex for this type of approach; it is then possible to analyze the spectrum in its entirety or for a specific spectral window, taking into account all of the vibration bands present. The exploitation of IR spectrometry data as a whole requires the use of multivariate statistics. Due to the complexity of the spectra, generated by the many components of food matrices, the use of complex calibration models is often required, based on the empirical study of the relationship between the concentration of compounds of interest (measured by conventional techniques) and the spectral measurements at different wave numbers [FIE 02]. The models calculated by PLS (partial least square) regressions, denoted PLS-R and proposed for the first time by Wold *et al.* in 1989 [WOL 89], were thus exploited to estimate the total lipid, carbohydrate and protein content as well as for the analysis of the dry matter content and other physico-chemical properties such as the sweetening power of many food matrices, such as milk, oils, fruit juices [VAN 92].

In this approach, the chemometric prediction model which is developed allows us to link a set of Y responses (the concentrations of the analytes of interest) as a function of a set of X variables (the spectrometric data) [ROM 06, TEN 95].

The theoretical infrared spectrum of the matrix studied is thus reconstructed from the infrared spectra of its main constituents (lipids, carbohydrates and proteins), taking into account the additivity relation of the Beer–Lambert law for each of these macromolecular contributions (equation [3.1]). It is thus possible to deduce the relationship between the shape of the spectrum of the matrix and the abundance of its constituent macromolecules.

$$D(\nu) = d\sum_{i=1}^{n} \varepsilon_i(\nu)c_i \qquad [3.1]$$

Equation [3.1] shows the additivity property of the Lambert–Beer law for a mixture of n chemical compounds. The optical density $D(\nu)$ of a mixture of compounds not undergoing chemical interaction is equal to the sum of the partial optical densities of its compounds at a given frequency ν and for a given molar extinction coefficient $\varepsilon(\nu)$. This is the concentration of the compound and the length of the optical path of the light in the sample.

For a multiple linear regression model to have the best possible reliability, it is necessary to perform a calibration that takes into account the range of linearity specific to each analyte, generate the corresponding model or models and finally, test their robustness by cross-validation and with a specific test set. This leads to statistical indicators such as RMSECV (Root Mean Square Error of Cross-Validation) or RMSEP (Root Mean Square Error of Prediction), which together with R^2 (coefficient of determination), are details on the quality of the model(s) generated.

Although PLS models are the most common [PIC 06], other multivariate statistical analyses are used for the study of food matrices, such as principal component regression or multiple linear regression [LIC 02].

Another method of statistical analysis of spectra involves comparing the infrared profiles of contrasting samples; this approach, which is also based on the use of complex calibration models, can make it possible to differentiate the spectra of products that have undergone physico-chemical modifications, or classify samples into different categories according to their profile. This approach uses univariate statistical tools, such as correlation analyses (Pearson, Spearman, etc.), or multivariate statistics, such as principal component analysis or PLS-DA (partial least square discriminant analysis) [GRE 14].

3.3.2.2. *Mid-infrared spectroscopy (MIR)*

MIR spectroscopy is commonly applied to the analysis of moisture, protein content, lipids and carbohydrates of food products in industries. Fourier Transform Infrared Spectroscopy (FTIR) is the most commonly used technique. FTIR spectroscopy allows analyses of both the structure–function relationship of the compounds studied and their quantification, making this technique a reference tool for research as well as quality control in the food industry [NAW 13].

3.3.2.2.1. Instrumentation

The first measuring instruments in the mid infrared operated on the principle of distortion spectrometers, based on the principle of transmission spectroscopy. With this technique, the sample is exposed to IR radiation and a detector analyzes the fraction of radiation absorbed at specific frequencies. A diffraction grating is used as a prism to separate the wavelengths of light into different spectral windows and send them separately to the detector through an aperture. The absorbances at each wavelength can then be analyzed one by one by varying the position of the diffraction grating and the aperture. The analysis of many wavelengths with this type of instrument is tedious, with significant analysis timescales, and thus limits the number of readings for each sample. The development of computers and spectroscopic techniques based on the Fourier transform has revolutionized the potential of infrared spectroscopy. In an FTIR device, the principle used is always that of transmission, but the way in which the spectrum is recorded is very different. The specificity of the FTIR is related to the use of an interferometer formed by an IR source, a beam splitter, two mirrors and a detector (Figure 3.4).

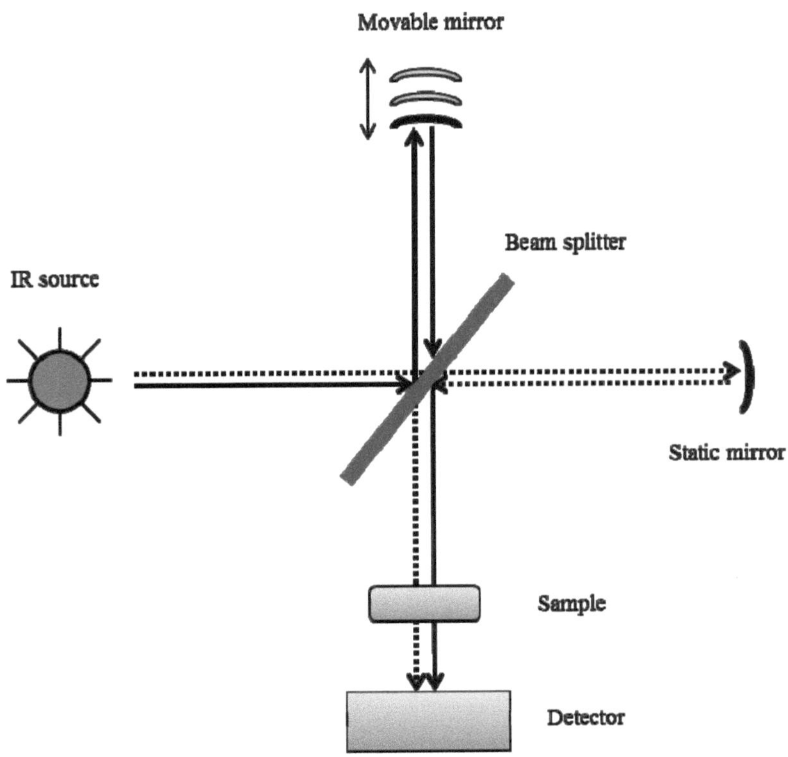

Figure 3.4. *Diagram of an interferometer as used in FTIR (original figure in [BLU 12])*

Here the IR radiation is separated and redirected to two mirrors, one static and one mobile. The radiation is reflected by the mirrors, is recombined by the beam splitter and then reaches the detector, which simultaneously measures the absorbances at all wavelengths. Once the two beams are recombined by the beam splitter, after following paths of different lengths, it is possible to measure constructive and destructive interferences which form what is called an interferogram. Here, the data measured by the detector corresponds to the Fourier transform of the spectrum, and so we apply another Fourier transform to this data in order to reconstruct the IR spectrum of the analyzed product. This technique allows a much faster analysis of the samples, and a greater amount of energy reaches the detector, which in turn allows an increase in the signal-to-noise ratio. Almost all of the spectrometers dedicated to the study of the spectral window of the mid infrared are currently based on the FTIR principle [BLU 12].

The use of FTIR therefore increases the potential of MIR spectroscopy for the study of complex matrices; however, as this technique is based on the transmission principle, the steps for preparing solid or liquid samples are the same as those for analysis by dispersive spectrometry. These time-consuming preparations also induce additional variability, depending on the operator who prepared the samples. Moreover, transmission MIR spectroscopy makes the analysis of matrices with high water contents difficult due to the high contribution of the latter to the spectrum, since water has predominant absorption bands on a spectrum obtained by FTIR (see section 3.3.2.2.2.); it is possible to limit this impact by using sampling systems that include short optical paths or by using D_2O instead of water.

The development of ATR (Atttenuated Total Reflection) has provided an alternative to the use of transmission spectroscopy. In this approach, the IR ray passes through a crystal with a high refraction index (diamond, germanium, etc.) at a specific angle, allowing the total reflection of the ray at the surface of the crystal in contact with the sample. This reflection causes an evanescence wave passing through the crystal to the area on which the sample is placed (Figure 3.5). The penetration of the radius in the sample is at a depth of between 0.5 and 5 µm. The absorbance induced by the sample will impact the evanescence wave, and so after one or more rebounds, the ray leaves the crystal and reaches the detector [BLU 12].

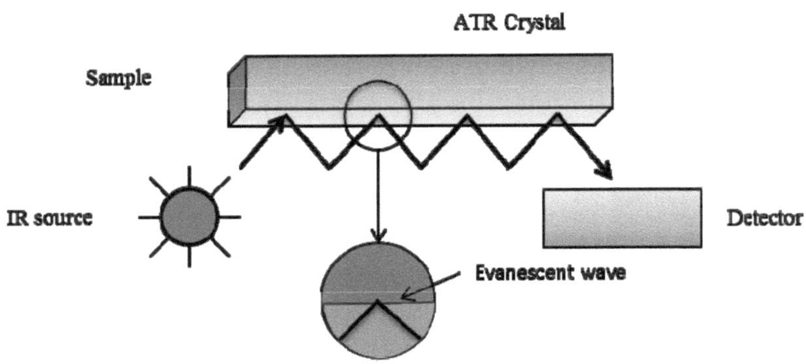

Figure 3.5. *Diagram of an ATR crystal with four reflections (original figure in [BLU 12])*

Regardless of the acquisition resolution scale, most of the bands found in an MIR spectrum derived from the analysis of a food product have been identified and assigned to particular vibrational modes corresponding to different chemical groups [KAR 03]. For example, the absorption bands of amides I (C=O elongation ≈ 1600–1690 cm^{-1}), II (C-N elongation and N-H deformation ≈ 1480–1575 cm^{-1}) and III (C-N elongation and N-H deformation ≈ 1230–1300 cm^{-1}) of proteins, that of lipids (C-H elongation ≈ 2800–3000 cm^{-1}, C-O elongation ≈ 1725–1745 cm^{-1}, C-O elongation of triacylglycerides ≈ 1175 cm^{-1}, C=C-H deformation ≈ 970 cm^{-1})

and finally that of carbohydrates (C-H elongation ≈ 2800–3000 cm^{-1}, elongation and deformation of the skeleton ≈ 800–1400 cm^{-1}). Water absorbs infrared radiation very well and has characteristic bands (H-O elongation ≈ 3360 cm^{-1}, water association band ≈ 2130 cm^{-1}, H-O-H deformation ≈ 1640 cm^{-1}) [NAW 13, KAR 06]. An example of an annotated FTIR spectrum of LE is presented below (Figure 3.6).

The main vibration bands of interest are indicated on the spectrum. Only the most informative region is represented. The bands Amide I and II (≈ 1600–1690 cm^{-1} and ≈ 1480–1575 cm^{-1}) correspond to the protein signal, the bands between 2800 and 3000 cm^{-1} and close to 1740 cm^{-1} correspond to the lipid signals (lipoproteins and triacylglycerols) and the bands between 1000 and 1200 cm^{-1} correspond to the carbohydrate signals (free glucose, osamines).

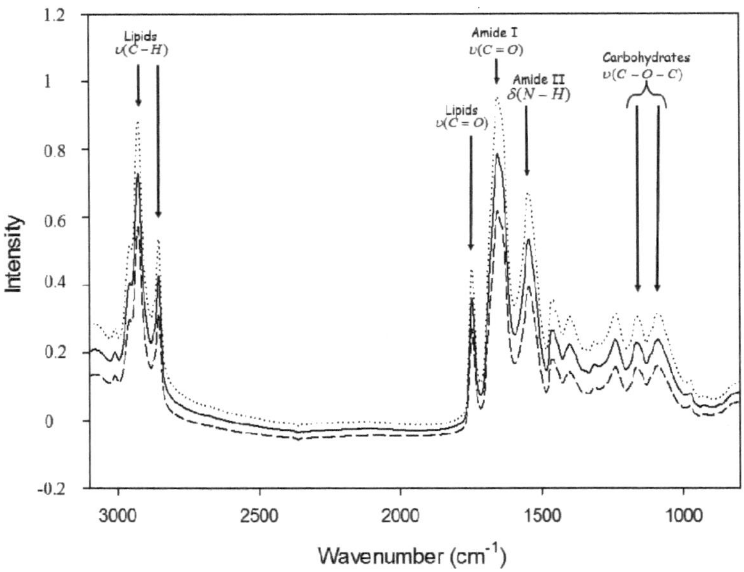

Figure 3.6. *FTIR spectrum recorded on a sterile liquid whole diluted to 1/10th and dried in a vacuum desiccator (5µL deposit). The experiments were carried out in the Process Engineering for Environment and Food laboratory (GEPEA – UMR CNRS 6144) at the University of Nantes, France*

3.3.2.2.2. Importance of signal aliasing

FTIR is also a powerful tool for the analysis of the secondary structure of proteins during the production or storage of food products [LIC 02]; however, these analyses need to take precautions with regard to signal aliasing problems. Indeed, the analyses relating to proteins, more particularly those on their secondary structure, are carried out via the study and the deconvolution of the amide region I (≈ 1700–1600 cm^{-1}); spectra resulting from FTIR analyses outside of this region are strongly impacted by the contribution of water (≈ 1640 cm^{-1}), as shown in Figure 3.7, extracted from a study on the characterization of the photosynthetic system of microalgae.

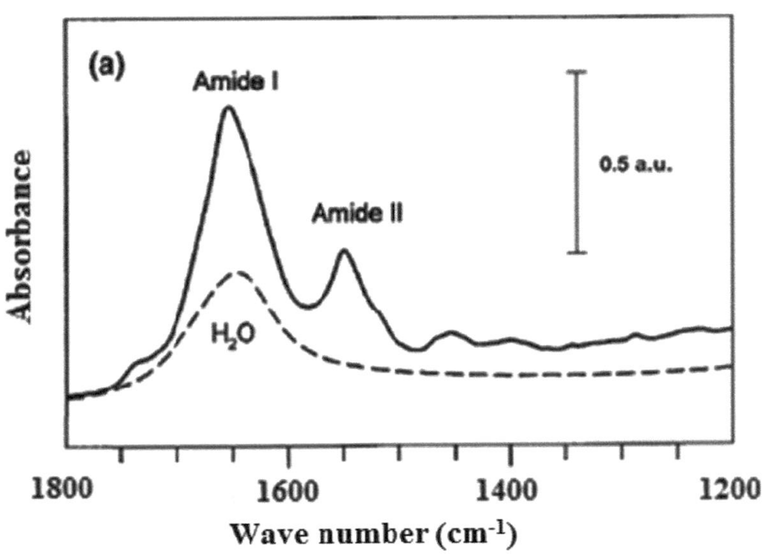

Figure 3.7. *Absorption spectrum of photosystem II in thylakoid membranes (continuous line) and the contribution of water molecules (discontinuous line) [BER 09]*

This remark is essential in cases where the spectral information is processed by a vibrator of interest (i.e. for a given wave number). There are several solutions by which to overcome this problem, that is, choosing the bands with the least possible aliasing, or when this is not possible, for example due to the complexity of the sample, resorting to more global processing methods, which will therefore exploit the influence of all of the spectral contributions for a given sample. For some analyses, it is also possible to perform the analysis in a D_2O solvent in order to overcome the contribution of water in this spectral region.

3.3.2.2.3. Application of FTIR to the study of LE spoilage

Infrared spectroscopy was first applied to the study of LEs in 1982 [LIT 82], inspired by the work carried out 20 years previously on the quantitative analysis of milk by this same technique [GOU 64]. This helped develop a technique with which to quantify dry matter, proteins and total lipids in liquid egg products. However, it is only with the development of computer science and spectrometric approaches based on the Fourier transform that it has been possible to efficiently process the information contained in these spectra, notably by exploiting the techniques derived from multivariate statistics. In particular, FTIR has been used to determine the lipid content and moisture content of food samples [VAN 92, VAN 93] and many teams have used this technique to study the secondary protein structure of eggs [SAN 88, FUR 07, BLU 15]; however, very few approaches have been developed to study the freshness of shell eggs [UNE 10] and none, to our knowledge, relate to the study of liquid whole egg spoilage.

Liquid whole egg contains a significant amount of water (88.5%) [KAR 09], and this water effectively absorbs infrared radiation, which generates a strong reading on the MIR spectrum. In fact, water has predominant absorption bands (H-O elongation $\approx$ 3360 cm^{-1}, water association band $\approx$ 2130 cm^{-1}, H-O-H deformation $\approx$ 1640 cm^{-1}) which can, because of their position and their intensity, mask signals of interest (see section 3.3.2.2.2). It is possible to limit this contribution by drying the samples through various methods (drying, lyophilization, IR heating, etc.) before reading, or to use specific correction algorithms designed to eliminate the contribution of water to IR spectra [POW 86, BOU 03]. The precise subtraction of water absorbance bands is possible due to the accuracy of frequency available in FTIR [KAR 03].

Figure 3.8 shows two FTIR spectra of vacuum-dried contrasted LE samples corresponding to the first and the fourteenth day of analysis in a kinetic study of the microbiological degradation of ELs.

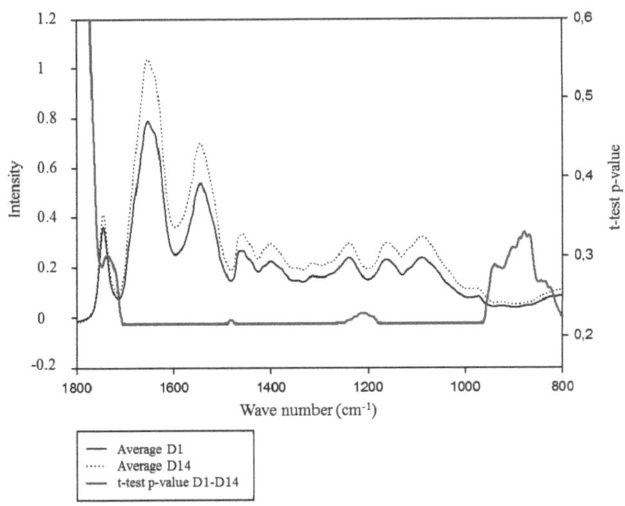

Figure 3.8. *FTIR spectrum of LE samples at days 1 and 14, incubated at 8°C after inoculation of a bacterial species representative of occasional spoilage of LEs in an industrial setting. The readings were carried out by transmission using an HTS-XT microplate reader (Bruker, France) on samples diluted in water at 1/10th and dried in a vacuum. The spectra presented for days 1 and 14 are averaged over nine replicas. The thick line corresponds to the p-value of a t-test carried out on the 18 samples. Only the area of interest is presented. The experiments were carried out in the Process Engineering for Environment and Food laboratory (GEPEA – UMR CNRS 6144) at the University of Nantes, France. For a color version of this figure, see www.iste.co.uk/ goncalves/ovoproducts.zip*

The two averaged spectra have very similar profiles; the p-value results of a t-test performed on the replicas (greater than 0.2 for all of the measured wave numbers and therefore not statistically significant) show that it is very difficult to differentiate between these spectra by an approach derived from univariate statistics.

Figure 3.9 presents the results of a multivariate analysis using PLS-DA (Partial Least Square Discriminant Analysis) performed on the data from the previously shown FTIR spectra. No standardization or filtration steps were performed on these data.

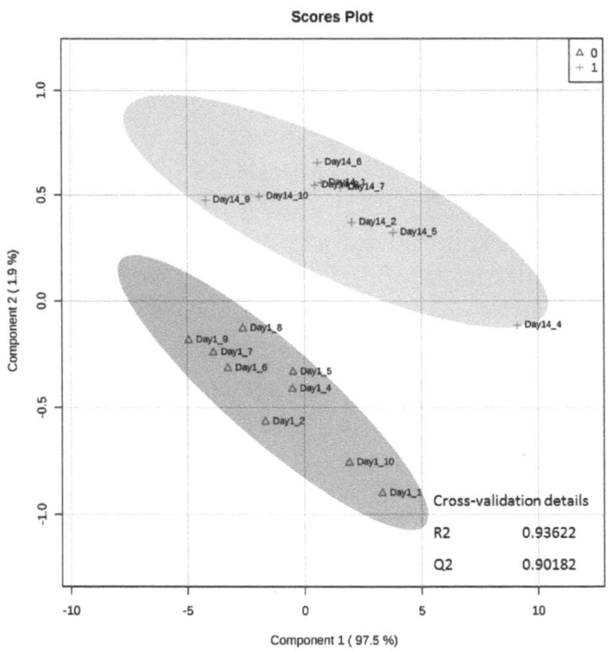

Figure 3.9. *PLSDA (Partial Least Square Discriminant Analysis) of data from FTIR analyses of LE samples at days 1 and 14, incubated at 8°C after inoculation of a bacterial species representative of occasional spoilage of LEs in an industrial environment. The experiments were carried out in the Process Engineering for Environment and Food laboratory (GEPEA – UMR CNRS 6144) at the University of Nantes, France. For a color version of this figure, see www.iste.co.uk/goncalves/ovoproducts.zip*

PLS-DA analysis here allows, without any preprocessing of the data, the effective separation of the two groups of samples according to their incubation time after inoculation, thus illustrating the potential of this technique for the analysis of LE spoilage.

By virtue of its ability to generate a "chemical profile" of the sample, FTIR is therefore a promising technique for the analysis of freshness and spoilage of LEs. However, the interpretation of these spectra requires the use of approaches derived from multivariate statistics and are therefore limited by the biases inherent in its data interpretation techniques, such as the effects of "overfitting" for example (see section 3.3.2.1.8).

3.3.2.3. *Near-infrared spectroscopy (NIRS)*

The potential of NIRS for the study of food products has been demonstrated since the 1960s [NOR 61], as many studies have confirmed the ability of this technique to quickly provide reliable information on the concentration of the main constituents of different food products [OSB 00]. Its ability to simultaneously quantify several compounds (total lipids, proteins, carbohydrates and moisture content) in different matrices in a very short time gives it a significant advantage over conventional analysis techniques, in addition to the fact that it equals or even surpasses them in terms of accuracy [NAW 13].

Another important advantage of NIRS is the ability to analyze many types of solid or liquid samples without or with just a few preparation steps. In particular, liquid samples can be analyzed in an automated way, and the use of optical fiber opens online monitoring possibilities. NIRS also allows the study of compounds with high water content without the need for drying, unlike MIR spectroscopy. The water signal is indeed much more attenuated in this spectral zone. These advantages are, however, counteracted by the need to perform specific calibrations of the instruments used and the matrices studied, and therefore to renew them systematically in the event of a change in the formulation of the products studied. NIRS also has a sensitivity that is often too low for the study of minority components [LIC 02]. Nevertheless, NIRS remains one of the most promising non-destructive analytical methods in the agri-food sector [LAN 11].

3.3.2.3.1. Instrumentation

The absorbance bands of the near infrared region are due to the harmonic frequencies and combinations of fundamental vibration patterns observed in the MIR region. Vibrations involving C-H, O-H, N-H bonds and ultimately S-H and C=O bonds, are responsible for the majority of the absorbance bands in the NIR region [LIC 02].

The frequency of the harmonics corresponds to approximately double, triple, etc. the frequencies of fundamental vibrations. The bands of combinations come from the interaction of two or more simultaneous vibrations, their frequency being the sum of the multiples of the frequencies of the corresponding fundamental vibrations [NAW 13].

The intensity of the bands corresponding to harmonics and association bands in NIR decreases rapidly while their frequency increases (the second harmonic will present a lower intensity than the first, etc.). As a result, the NIR region is characterized by low intensity bands, generally between 10 and 100 times lower than those of the MIR region. The highest harmonics with a lower intensity, only the signals whose fundamental vibration is located in the spectral region of the MIR (between 2.7 and 5.4 μm), will present a relatively high intensity in the region of the NIR (the first harmonic of vibrational modes in question will be present in the NIR region) [CHA 84].

The first "near-infrared" spectrometers differed from UV-visible spectroscopy instruments only by minor modifications. The potential of these devices has led to the development of equipment specific to NIRS. First, devices used scanning monochromators and large aperture holographic gratings. The scanning spectrometers require significant signal-to-noise ratios to be functional, thus implying a meticulous electronic design. These highly versatile spectrometers required an advanced understanding of the principle of NIRS, limiting their application in an industrial environment. For routine applications, simpler systems based on the use of filter sets replacing the monochromator have been developed. These devices allow the user to simply perform measurements at different predefined wavelengths [CHA 84].

As with MIR spectroscopy, NIRS has benefited greatly from the development of microcomputer and spectroscopic approaches based on the Fourier transform, thus giving rise to the FT-NIR. This technique, which works on the same principle as FTIR, has made it possible to improve the analysis speed, the reproducibility of the spectra and the accuracy of the measured wavelengths.

Spectrometers for analyzing the NIR region operate on the principle of transmission, diffuse reflection or a mixture of both, referred to as "transflectance" (Figure 3.10). These various techniques have provided NIRS with a wide range of sampling methods and optical probes for the analysis of numerous matrices without sample preparation, and they point toward possible online analysis.

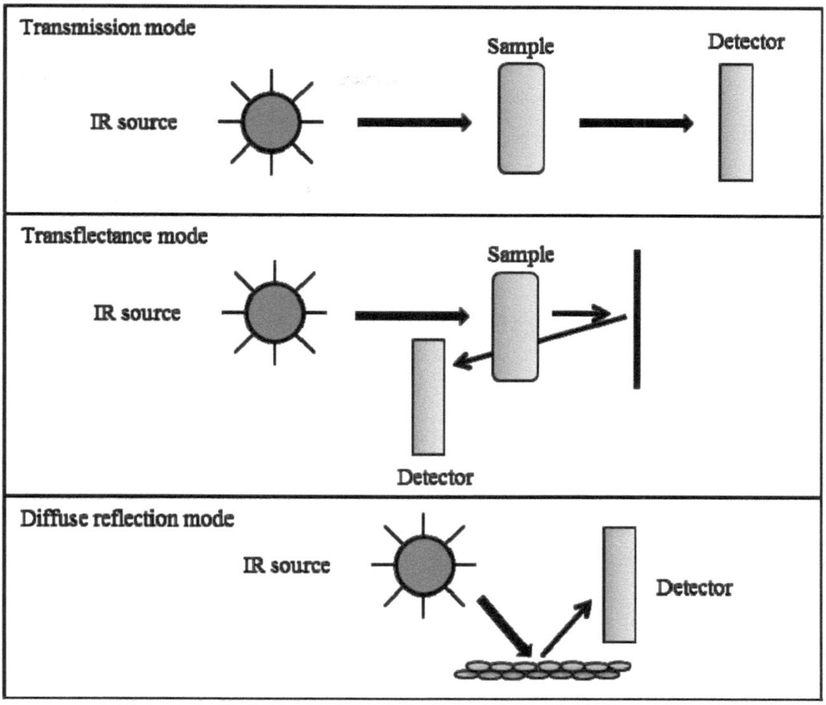

Figure 3.10. *Diagram of the transmission, "transflectance" and diffuse reflection modes (inspired by the original figure in [HAN 11])*

Recent advances, such as near-infrared microscopy [DEL 07] and near-infrared hyper-spectral imaging systems [FER 12], have made it possible to perform NIR analyses at the macro or microscopic scale as well as to determine new physicochemical parameters such as the presence of omega-3 fatty acid in food matrices.

The main vibrational bands in the NIR domain used in the quantitative analysis of major food components are presented in Table 3.7.

Component	Wavelength (nm)	Wave number (cm^{-1})	Assignment
Proteins	910	10989	C–H elongation, third harmonic
Lipids	928	10776	C–H elongation, third harmonic
Water	970	10309	O–H elongation, second harmonic
Starch	990	10101	O–H elongation, second harmonic
Proteins	1020	9804	2*N–H elongation + 2*amide I
Lipids	1037	9643	2*C–H elongation + 2*C–H deformation + (CH2)n
Lipids	1200	8333	C–H elongation, second harmonic (CH2 groups)
Sucrose, Starch	1440	6944	O–H elongation, first harmonic
Water, Starch	1450	6897	O–H elongation, first harmonic
Glucose	1480	6757	O–H elongation, first harmonic (intramolecular hydrogen bond)
Cellulose	1490	6711	O–H elongation, first harmonic (intramolecular hydrogen bond)
Proteins	1510	6623	N–H elongation, second harmonic
Starch	1528	6545	O–H elongation, first harmonic (intramolecular hydrogen bond)
Starch	1540	6494	O–H elongation, first harmonic (intramolecular hydrogen bond)
Starch, glucose	1580	6329	O–H elongation, first harmonic (intramolecular hydrogen bond)
Lipids	1734	5767	C–H elongation, first harmonic (CH2 groups)
Lipids	1765	5666	C–H elongation, first harmonic (CH2 groups)

Component	Wavelength (nm)	Wave number (cm^{-1})	Assignment
Cellulose	1780	5618	C–H elongation, first harmonic
Cellulose	1820	5495	O–H elongation + 2*C–O elongation
Starch	1900	5263	O–H elongation + 2*C–O elongation
Water	1940	5155	O–H elongation + O–H deformation
Proteins	1980	5051	N–H asymmetric elongation + amide II
Starch	2000	5000	2*O–H deformation + C–O deformation
Proteins	2050	4878	N–H symmetrical elongation + amide II
Sucrose, Starch	2080	4808	O–H elongation + O–H deformation
Starch	2100	4762	2*O–H deformation + 2*C–O elongation
Proteins	2180	4587	2*amide I + amide III
Starch	2252	4440	O–H elongation + O–H deformation
Starch	2276	4394	O–H elongation + C–C elongation
Cellulose	2336	4281	C–H elongation + C–H deformation
Cellulose	2352	4252	C–H deformation, second harmonic
Starch	2461	4063	C–H elongation + C–C elongation
Starch	2488	4019	C–H elongation + C–C elongation
Starch	2500	4000	C–H elongation + C–C elongation

Table 3.7. *Vibrational bands in the NIR domain used in the quantitative analysis of major food products (data according to [LIC 02], adapted according to work by [OSB 93])*

3.3.2.3.2. Importance of signal aliasing

The absorbance bands located in the region of the NIR are wide and numerous, which generates a large number of aliasing signals on the spectra. These constraints require a more complex interpretation and have therefore limited the number of basic studies on food products. NIRS was therefore used mainly for the quantitative analysis of the main food components (lipids, proteins, total sugars and moisture content) and to a lesser extent the qualitative analysis (identification, discrimination) of the components of these same products [LIC 02].

The complexity of food spectra in NIRS involves the use of complex calibration models that employ the entire spectrum or at least several wavelengths (see section 3.3.2.1.8).

3.3.2.3.3. Application of NIRS to the study of LE spoilage

The first applications of NIRS for the analysis of egg products date back to the 1980s and concerned the measurement of dry matter. Approaches for measuring moisture content and total protein and lipid contents were subsequently developed fairly quickly [WEH 88]. The purpose of these studies was to evaluate the applicability of a commercially available NIR spectrometer to the simultaneous study of moisture content and total protein and total lipid content of dried egg product samples by atomization.

Different NIRS approaches using FT-NIRS (Fourier transform near-infrared spectroscopy) have recently been developed for the non-invasive analysis of the quality and freshness of shell eggs [ZHA 10, ZIU 08, LIU 11]. The results of these analyses are generally compared to internationally recognized criteria for the freshness of shell eggs (Haugh unit, albumen pH, etc.). Their interpretation is done through linear or non-linear multivariate statistical models.

NIRS was also used to study changes in the physico-chemical composition of freeze-dried egg yolks (especially polyunsaturated fatty acid composition) according to the egg-layer diet [DAL 06]. In this study, PLS models allowed the almost perfect separation of samples according to the different diets of the egg-layers (different methods of polyunsaturated fatty acid enrichment); however, other parameters (such as pH, cholesterol content and CIE color parameters) could not be correctly predicted.

NIRS is therefore used in many applications in the field of egg and egg product analysis, and although several studies have explored the ability of this method to assess the freshness of liquid whole eggs, none to our knowledge directly relate to their spoilage in the production environment. Besides the ability of this method to generate a "physico-chemical signature" of the sample (just like MIR spectroscopy), it is a potentially interesting candidate for the study of this spoilage.

3.3.3. *Raman spectroscopy*

3.3.3.1. *Overview*

Raman scattering was considered theoretically in the mid-1920s. Several research teams were working on this subject at the same time, but the experimental evidence of this effect was first described by Chandrasekhara Venkata Raman [RAM 28]. After this discovery, most of the experimental aspects were quickly explored (63 publications appeared on this subject in 1928 and 141 in 1929). At the end of the 1930s, vibrational spectroscopy (Raman and infrared) became the main non-destructive analytical methods, but Raman spectroscopy still required very skilled operators, since Raman scattering is a very weak phenomenon and thus requires the use of sources of bright lights. The most important applications emerged following the combination of lasers in Raman spectrometers in the 1960s [KOG 63]. Despite the development of these new sources of irradiation, Raman spectroscopy remained largely confined to laboratories as it always required qualified teams and very specific conditions in order to obtain exploitable spectra. Reliable analyses and bibliographic data were not at all widespread, and Raman spectroscopy was strongly affected by the fluorescence phenomenon now overcome by the use of lasers in the red or the near infrared [DIE 93]. This excitation significantly reduced the fluorescence, which made the use of powerful lasers possible without causing the photo-decomposition of the sample.

3.3.3.2. *Principle*

3.3.3.2.1. Conventional Raman effect (linear)

Raman scattering spectroscopy, like infrared absorption spectroscopy, is a technique used to observe the vibrational states (also rotational in some cases) of a molecule. It makes it possible to characterize the molecular composition of a sample by identifying the chemical groups that constitute it. This technique is complementary to infrared spectroscopy, which also makes it possible to study the natural vibrational modes of molecules, but it is distinguished by the ability to perform measurements in aqueous media, since infrared spectroscopy is very sensitive to polar media. To observe the Raman effect, we exploit the interaction between a sample and monochromatic radiation, that is to say, an electromagnetic radiation whose frequency is accurately determined. A laser source is generally used for this purpose. Some of the photons forming the radiation are either reflected or absorbed, and few are scattered. Most of these photons are of the same frequency ν_0 as the incident radiation. This is a scattering phenomenon without frequency change, called Rayleigh scattering or elastic scattering (approximately one photon in a thousand is elastically scattered). It is estimated that only one in 10^8 photons are diffused inelastically with a change of frequency, and this is what constitutes the Raman effect. The frequency of scattered photons is called ν_d. There are two cases of this:

If $\nu_d > \nu_0$, then $\nu_d = \nu_0 - \nu_v$: stokes Raman scattering

If $\nu_d < \nu_0$, then $\nu_d = \nu_0 + \nu_v$: anti-stokes Raman scattering.

In these two cases, the frequency differences ν are equal to the vibration frequencies, active in Raman, for the studied compound.

An energy representation of the Raman scattering process is presented in Figure 3.11. The energetic state of any molecule can be described from the rotational, vibrational and electronic wave functions associated with it. The energy E of the system is equal to the sum of the three energies: $E = E_{rot} + E_{vib} + E_{el}$; these three energies have increasing values, on several orders of magnitude. E0 and E1 respectively represent the fundamental and excited electronic level of a molecule, whereas the discrete levels “v” represent the vibrational levels. During the scattering process, if the system returns to its starting level, it is called Rayleigh elastic scattering. If, on the other hand, the system stabilizes on a vibrational level different to the initial level, the scattered photon will have a reduced or increased energy of precisely $E_{v1} - E_{v0}$. This energy difference is necessary for a molecule of interest to vibrate. As an indication, the equivalent transition measured by infrared absorption spectroscopy is presented on this same graph. Fluorescence and phosphorescence emission processes can also occur and often result in intense and very wide bands that obscure and obstruct the Raman spectrum. Other Raman effects are added to the classic Raman effect (the most widely studied), [SCH 97]. We highlight the use of Surface-Enhanced Raman

Scattering (SERS) spectroscopy [CHE 82, MUR 82]. The electromagnetic aspect of this phenomenon is based on strong electromagnetic confinement effects caused by the interaction of the molecules of the sample with metallic surfaces, primarily gold or silver [VOD 98]. The compounds studied must be near-intracellular metal aggregates (nanometric colloïds of silver or gold) or extracellular (rough or nanostructured surfaces). The resulting exaltation of the field causes an increase in the Raman signal by several orders of magnitude, which allows the spectra to be recorded with a very high sensitivity. This sensitivity depends on the chemical nature of the adsorbed molecule and the type of metal surface used. It differs only from conventional Raman by the specific preparation of samples and not by the apparatus [OTT 91, KUD 05, ARO 06]. Despite its effectiveness in exalting the Raman signal, this technique has some drawbacks, such as selective enhancement of certain Raman bands and poor reproducibility of the spectra obtained.

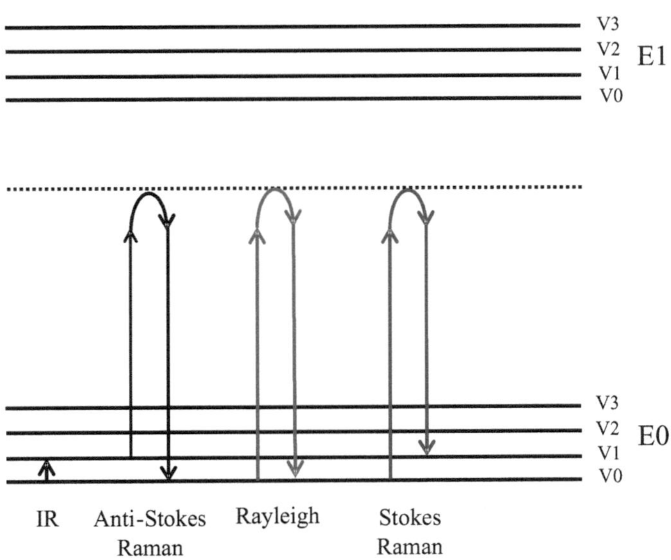

Figure 3.11. *Jablonski diagram indicating the Raman vibrational modes during excitation by a laser source (E = energy level, V = vibrational level)*

Resonance Raman scattering occurs when the wavelength of the exciting radiation is close to that of an electronic transition of the analyzed compound. The spectra obtained contain precise information on the base state of the molecule, as well as on its excited state. Raman resonance scattering is characterized by a very strong exaltation of the intensity of certain peaks (up to a factor of 10^6) of the chromophores involved in the exalted electronic transition.

3.3.3.2.2. Unconventional Raman effect (non-linear)

Other non-linear effects are added to these Raman effects, which are already mentioned. They are measurable when the electric field created by the incident source is very strong. This is the case of pulsed lasers that can obtain around one gigawatt of power in a very short space of time (10^{-12} s). These nonlinear effects allowed the development of approaches such as Hyper-Raman Scattering, Stimulated Raman Gains Spectroscopy (SRGS), Inverse Raman Spectroscopy (IRS) or Stimulated Raman Loss Spectroscopy (SRLS), Coherent Stokes Raman Spectroscopy (CSRS), Coherent Anti-stokes Raman Spectroscopy (CARS) and Photoacoustic Raman Spectroscopy (PARS).

3.3.3.3. *Instrumentation*

The main elements of a Raman spectrometer are shown schematically in Figure 3.12. Their roles can be summarized below:

– the laser provides the source of monochromatic irradiation; the Raman spectrometer can be equipped with several lasers of different wavelengths chosen according to the desired application or to avoid the weak Raman signal being masked by strong luminescence. Figure 3.12 shows the wavelengths of the most-used lasers in Raman spectroscopy.

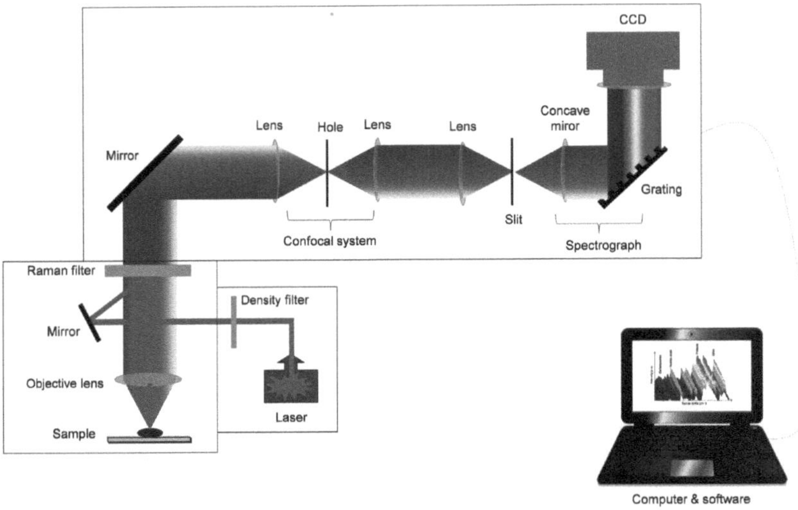

Figure 3.12. *Principle of a Raman spectrometer (confocal). For a color version of this figure, see www.iste.co.uk/goncalves/ovoproducts.zip*

Ultra-Violet	Visible	Infrared
244 nm (Argon)	488 nm (Ar^+ : Blue)	785 nm (Diode Laser)
325 nm (He-Cd)	514 nm (Ar^+ : Green)	830 nm (Diode Laser)
363 nm (Ar^+)	532 nm (YAG : Green)	
	633 nm (He-Ne : Red)	

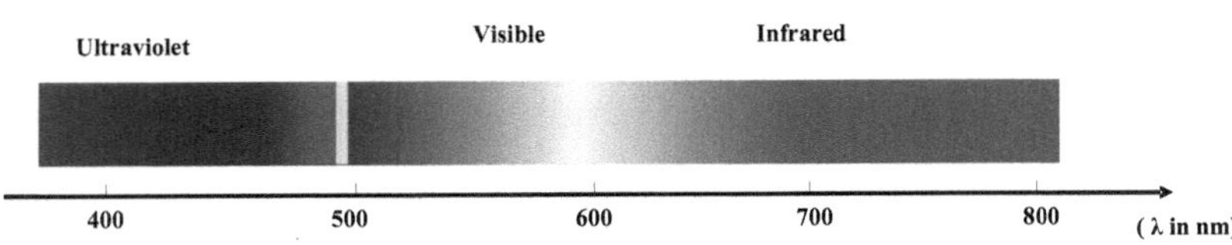

Figure 3.13. *List of wavelengths of the most-used lasers in Raman spectroscopy. For a color version of this figure, see www.iste.co.uk/goncalves/ovoproducts.zip*

– different types of filters (interferential, edge and density filters) can be integrated into the spectrometer, each with a well-defined role: (1) the interference filter located just after the irradiation source eliminates the parasitic radiation emitted by the laser, thus making it possible to improve the quality of the excitation beam; (2) the edge rejection filter eliminates the contribution of the Rayleigh scattering, which is often very intense compared to the Raman signal; (3) the density filter provides the user with the ability to control the power of the laser according to the analysis requirement.

– a focusing device on the sample to be analyzed is also used. In so-called macroscopic configuration, a single convergent lens is sufficient. However, regarding the application in a biological context, it is desirable to use a microscope. The microscope focuses the laser beam on the sample using different lenses and then collects scattered photons and redirects them to the spectrometer. The microscope also allows the operator through appropriate objectives to analyze very small sample quantities. In this case the analysis is done by backscatter.

– the confocal system makes it possible to choose the axial resolution thanks to the addition of a diaphragm with an adjustable diameter associated with the objective of the microscope. This, which does not exist on all spectrometers, makes it possible to considerably limit the analysis volume and thus improves the specificity of the recording.

– the spectrometer diffracts the received photons according to their wavelengths and then directs them to the detector (CCD camera). This part generally consists of one or more optical diffracting gratings, with the number of features defining the resolution of the apparatus.

– the CCD (Charge Coupled Device) detector corresponds to a multichannel detection and allows simultaneous collection of spatial and spectral information on the sample analyzed.

– the computer system transforms the electrical signal into a signal that can be used by the user, making it possible to control the device with different commands and the data processing with the dedicated software.

3.3.3.4. *The main characteristic bands of a Raman spectrum*

As we have already seen, Raman scattering is a vibrational characterization method in which the spectrum obtained corresponds to the decomposition of the electromagnetic radiation according to its different elements (the associated wavelengths). In order to compare spectra, it is advisable to use, not the radiation frequency, but a quantity which is proportional to it, the wave number. The wave number unit, cm^{-1}, is also called Kayser. Raman spectroscopy therefore uses the relative scale in cm^{-1} where the laser wavelength (λ_{laser}) corresponds to 0 cm^{-1} according to the following relation:

$$\text{Relative Raman wave number}\left(cm^{-1}\right)\times 10^{7} = \frac{1}{\lambda_{laser}(nm)} - \frac{1}{\lambda(nm)}$$

with λ = wavelength in nanometers.

When we use this scale in cm^{-1}, Raman spectra are always shifted to the same position. Standards for the presentation of Raman spectra have been proposed by the International Union of Pure and Applied Chemistry [SCH 97]. Thus standardized, the spectrum can then be considered as the fingerprint of the molecules that make up the analyzed sample. It provides various information; each line or band can be characterized by:

– its position in the spectrum, which can be related to the frequency of a specific mode of vibration, and the different lines of the spectrum can therefore be associated with chemical functional groups;

– its intensity, related to the number of scattered molecules as well as the mode of vibration considered;

– its state of polarization, which gives information on the symmetry of the chemical groups.

Table 3.8 presents the allocation of main Raman bands at different wavelengths.

Raman shifts (cm^{-1})	Assignment
497, 550	Carbohydrates : C -O-C
657	Tyrosine
720	Adenine
779, 810	Nucleic acids
850, 878	Tyrosine (in proteins)
1000	Phenylalanine
1093	Carbohydrates : (C-C, C-O)
1117	Unsaturated lipids : (=C-C=)
1243	Amide III : N -H, C-N
1311	Amide III
1444	$\delta(-CH_2)$
1568	Guanine, Uracil (cycle stretching)
1655	Amide I
2880	Lipids : CH_2 (symmetric stretching)
2933	Lipids : CH_3 (asymmetric stretching)

Table 3.8. *Assignment of the main Raman bands according to [SCH 00, MAQ 02]*

3.3.3.5. *Fields of application of Raman spectroscopy*

The field of application of Raman spectrometry is very broad and extends from physics–chemistry to the analysis of microscopic (Figure 3.14) or, more recently, nanometric biological species, for example the analysis of viruses [SHA 06]. It is thus possible, with very little sample preparation, to obtain information on the molecules in a living cell.

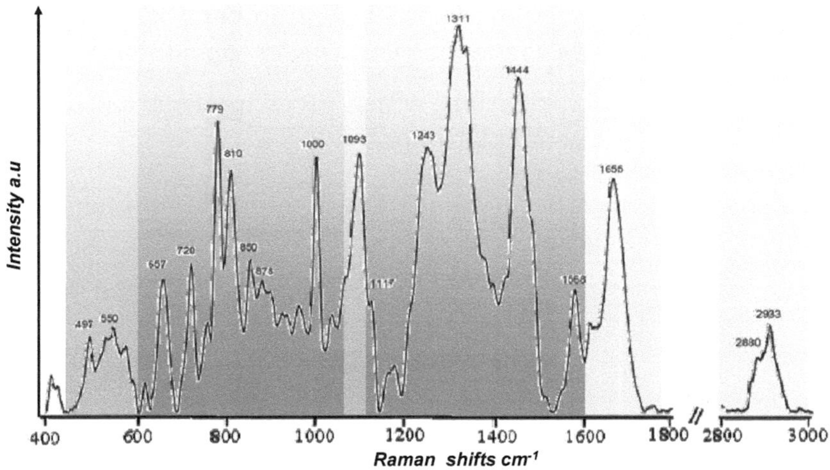

Figure 3.14. *Raman spectrum of a bacterium (*Salmonella enterica typhimurium*), recorded for the whole cell*

Raman spectrometry is also used for the quantitative determination of active substances and/or the homogeneity of the dispersion of the active agent in medicines, thus making it possible to highlight counterfeits [VAN 02]. The Raman effect has been exploited for the rapid identification of different kinds of drugs such as cocaine, heroin and phenylethylamine derivatives [BEL 00, RYD 99]. In medical diagnostics, Raman spectroscopy is widely used on whole cells and tissues. Thus, tumor cells can be searched for using fiber-optic probes to identify tumors in breast tissue [KNE 03, KRI 05]. In other applications, Raman spectroscopy has been very useful in assessing the exposure of human cells and tissues to toxic products or in characterizing the interaction between cells and toxins [NOT 04].

The use of Raman spectrometry in the agri-food sector is uncommon or non-existent, despite the latest technological developments that make it possible to obtain discriminative and versatile information on the very complex egg matrices, as shown in Figure 3.15.

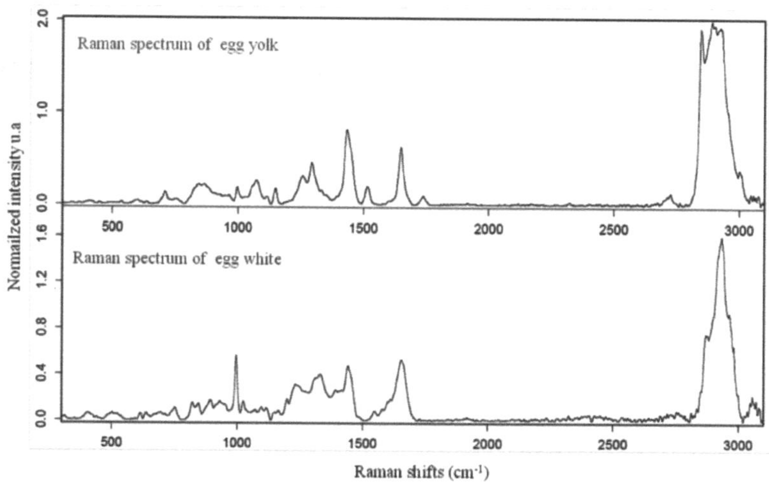

Figure 3.15. *Raman spectra of egg white and yolk recorded at 785 nm*

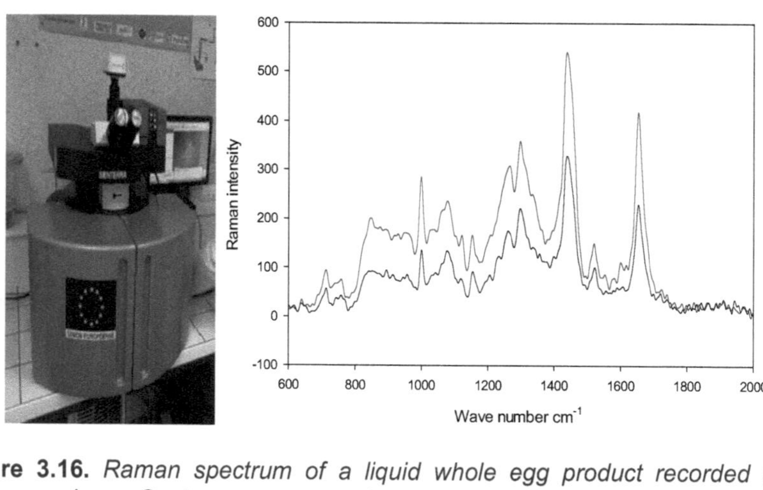

Figure 3.16. *Raman spectrum of a liquid whole egg product recorded at 532 nm using a Senterra compact microscope (Bruker) (left). On the right, superposition of two Raman spectra for two modes of preparation: fresh EP without any preparation (blue), and freeze-dried EP (black). No difference is to be noted, suggesting that the influence of water is almost negligible on the quality of the measured data. The experiments were carried out in the Process Engineering for Environment and Food laboratory (GEPEA – UMR CNRS 6144) at the University of Nantes, France. For a color version of this figure, see www.iste.co.uk/goncalves/ovoproducts.zip*

The chemical signatures recorded for the egg whites and yolks are clearly different and propose a wide application range for the control of egg products. Figure 3.16 shows another Raman signature, this time recorded at 532 nm and directly on an egg product (liquid whole egg), confirming the feasibility of these vibrational approaches.

Ultimately, these non-invasive approaches could replace conventional techniques which are often based on simple physico-chemical measurements (pH, conductivity, ash or water levels) or more complex analyses such as chromatography (HPLC, GC) or advanced spectrometric analyses (ICP MS). These often require rather onerous sample preparation or a wide range of commercial chemical standards for quantitative analyses. The lack of a database dedicated to agri-food matrices, essential for the rapid and robust interpretation of whole Raman spectra, is one of the main obstacles that may currently explain the low generalization of Raman. Efforts must therefore be made in this area.

3.4. Nuclear Magnetic Resonance for the study of the quality of egg products

This part aims to introduce the potential of *Nuclear Magnetic Resonance* (NMR) for the study of egg products. NMR is an analytical technique that is widely used in many scientific fields, from chemistry to biology, medicine or the study of materials. NMR and its many variations (relaxometry, imaging) are also widely used in the agri-food industry, but their applications in the field of egg products remains quite restricted in comparison with other analytical techniques. The aim of this chapter is therefore to make the reader aware of the possibilities opened up by NMR in this field.

3.4.1. *Introduction to Nuclear Magnetic Resonance*

NMR is a complex technique whose principle and implementation is based on notions of quantum physics that are outside of the scope of this book. The purpose of this section is to outline the main aspects of this technique, emphasizing the diversity of the techniques and associated equipment, as well as the advantages and limitations of NMR for the analysis of complex matrices such as egg products. For a better understanding of the NMR principle, many bibliographic references [ERN 87, CAN 02, KEE 10] or online courses offer a comprehensive description of this technique.

3.4.1.1. *Principle of NMR*

Besides being an analytical technique, NMR is a physical phenomenon based on the magnetic behavior of the atomic nucleus. The discovery of this magnetic behavior earned Isaac Rabi the Nobel Prize in Physics in 1944, while the discovery of the NMR phenomenon earned Felix Bloch and Edward Mills Purcell the Nobel Prize in Physics in 1952. In a very simplified way, the nucleus of an atom can be likened to a positively charged sphere. The rotational movement of this sphere can be seen as equivalent to an electric current producing a magnetic field, and the nucleus can thus be likened to a microscopic magnet with a north pole and a south pole. This characteristic of the atomic nucleus is called the spin magnetic moment. A sample contains a large number of nuclei whose magnetic fields are oriented in a totally random manner, with the resulting magnetic field of the overall sample being zero, which is why this magnetism is not observed at the macroscopic scale.

In an NMR experiment, the sample to be studied is placed in a very intense magnetic field (several tens of thousands of times that of Earth's magnetic field) denoted B_0. Under the effect of this intense field, the microscopic magnets – the nuclei – tend to align in the direction of the external magnetic field (as the needle of a compass is aligned with the direction of the Earth's magnetic field). In addition, this orientation is quantified, and the nuclei can take only two orientations, one in the direction of the magnetic field, and the other in the opposite direction (Figure 3.17a).

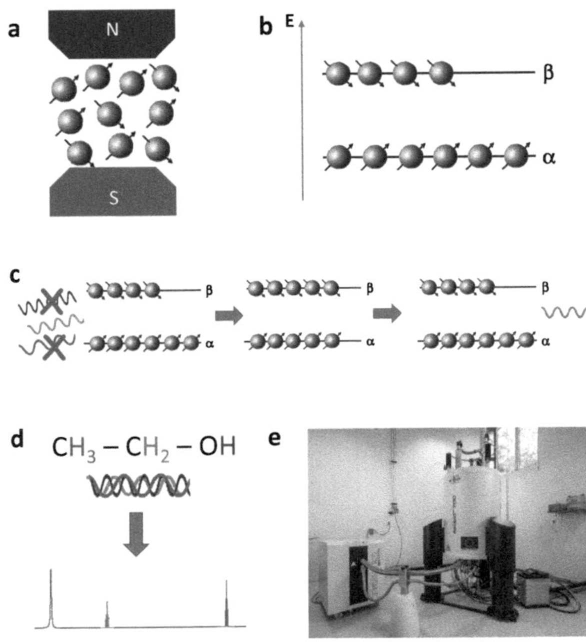

Figure 3.17. *Illustration of the principle of NMR. The nuclei of atoms placed in a magnetic field are oriented in a quantified manner (a), and this orientation results in a degeneracy of energy levels (b). The NMR experiment (c) involves an energy transition due to a radiofrequency wave (a single wave is effective; this is a phenomenon of resonance). The emission of this wave equalizes the populations; when the wave stops, the system returns to its equilibrium state by emitting a signal called Free Induction Decay (FID). This signal consists of a superposition of frequencies that characterize the chemical environment of the nuclei constituting the sample. In NMR spectroscopy, the Fourier transformation of this signal yields a spectrum (example of ethanol) whose analysis makes it possible to determine the structure of the molecules contained in the sample (d). The NMR spectroscopy experiment is carried out using an NMR spectrometer (e) (example of a high-field apparatus at 16.4 Tesla in the laboratory of Chemistry and Interdisciplinarity: Synthesis, Analysis, Modeling, at the University of Nantes, France). For a color version of this figure, see www.iste.co.uk/goncalves/ovoproducts.zip*

This orientation of the nuclei is reflected by the existence of two energy levels, called the *Zeeman effect* (Figure 3.17(b)). The lowest (and therefore the most stable) corresponds to the nuclei oriented in the direction of the magnetic field, while the highest level corresponds to the nuclei oriented in the opposite direction. Since the lower level (state α) is more stable, it is more populated than the upper level (state β), but the population difference at ambient temperature is very low (about 10^{-5}) because the difference between the energy levels is low with regard to thermal agitation. The NMR experiment is based on an energy transition between these two levels.

The NMR experiment involves sending an electromagnetic wave (whose magnetic field is denoted B_1) towards the sample (Figure 3.17c), which makes it possible to pass nuclei from the α state to the β state in order to achieve balanced nuclei populations on both levels. A single frequency is effective in causing this energy transition, which is why it is a resonance phenomenon.

When the electromagnetic wave stops, the excited nuclei return to their initial state, emitting a wave whose frequency depends on the difference between the energy levels. This wave is detected and forms the Nuclear Magnetic Resonance signal. This signal is called Free Induction Decay (FID).

The frequency of the NMR signal depends on the electronic environment of the nuclei, and therefore on their chemical environment. In one molecule, nuclei with a different chemical environment will give rise to distinct frequency signals. The FID is therefore the sum of signals evolving at different frequencies.

3.4.1.2. *Diversity of techniques and equipment*

The NMR phenomenon can be observed for all non-zero spin nuclei, that is, all those with magnetic behavior. Most nuclei in the periodic table of elements have at least one stable isotope with non-zero spin and can therefore be studied by NMR. The nucleus most commonly studied by NMR is hydrogen, and in particular its most abundant isotope, ^{1}H, whose nucleus is a simple proton. This is known as *proton NMR*. Other nuclei present in organic molecules, such as carbon-13 or nitrogen-15 are frequently studied using NMR. Their detection, however, is complicated by their low natural abundance (for example, for carbon, the most abundant isotope, ^{12}C, does not have a magnetic moment and is invisible in NMR). In this chapter, we will mainly focus on proton NMR.

There are different ways in which to exploit the signal detected in NMR, which has given rise to complementary techniques, the potential of which will be illustrated later.

3.4.1.2.1. NMR spectroscopy

NMR spectrometry involves performing a mathematical transformation of the FID, called Fourier transformation (FT), which separates the different frequencies comprising the FID. This operation (Figure 3.17(d)) makes it possible to obtain a spectrum, showing peaks on a relative frequency axis called the chemical shift axis (unitless quantity expressed in parts per million, or ppm). The analysis of this spectrum (position, area, fine structure of the peaks) makes it possible to determine the structure of the molecules constituting the sample, as well as to obtain information on the conformation of the molecules or their dynamics in solution.

3.4.1.2.2. NMR relaxometry

NMR relaxometry directly studies the FID resulting from the NMR experiment. The way in which this signal decreases characterizes the rate at which the nuclei return to their equilibrium state. This phenomenon, known as relaxation, provides valuable insight into the molecular dynamics and microscopic environment of nuclei.

3.4.1.2.3. Magnetic resonance imaging (MRI)

Magnetic resonance imaging (MRI) uses a specific-excitation approach to obtain a signal whose frequency depends on the spatial location of the nuclei. MRI requires the application of non-uniform magnetic fields (gradients). A specific processing of the resulting signal makes it possible to obtain macroscopic images of the samples studied in two or three dimensions. MRI, invented by Paul Lauterbur and Peter Mansfield, who obtained the Nobel Prize in Medicine in 2003, has quickly established itself as a major diagnostic technique in medicine. It is particularly used for soft tissue imaging (central nervous system, muscles, organs, tumors etc.). It is also widely used for the characterization of agri-food products.

3.4.1.2.4. Conclusion

This wide variety of methods makes it possible to study many types of samples using NMR. NMR spectroscopy can be applied to solid, liquid or even gaseous samples. It is the analytical tool of choice that chemists use every day to determine the structure of the molecules or materials that they synthesize. Thus, most research laboratories or large chemical companies have one or more NMR spectrometers for the study of liquid or solid samples. NMR is also a tool of choice for studying the structure and dynamics of biological molecules in the liquid or solid state, including proteins. In the case of purified molecules solubilized in liquid phase, the expert can directly analyze the spectrum using structural elucidation methods to determine the structure of the molecule in solution. In the case of more complex samples (extracts, biofluids, etc.), NMR analysis is often accompanied by statistical approaches to data processing. The study of solid samples by NMR spectroscopy is also widespread for the study of materials and biomaterials. Relaxometry and imaging make it possible to study samples of complex composition, comprising both solid and liquid phases. These techniques find many applications in life sciences, both for the study of samples of agri-food interest and for medical diagnosis.

3.4.1.3. *Benefits and limitations of NMR for the study of egg products*

NMR is a particularly suitable tool for the study of complex matrices such as egg products. NMR spectroscopy can be used to study the liquid whole egg, in the form of fractions (egg white or yolk) or in the form of extracts, making it possible to select certain families of molecules (metabolites, lipids, proteins) using a prior extraction step. Relaxometry and imaging, on the other hand, are especially adapted to the study of the whole egg.

NMR techniques have many advantages over other analytical techniques. First, NMR combines powerful *structural elucidation* capabilities with high *quantitative* potential. Indeed, the detected signal is directly proportional to the number of nuclei that give rise to the signal, and therefore to the concentration of the molecules in solution. The use of a reference method thus allows multiple titration of molecules in a mixture. Quantification can target certain molecules or classes of molecules, but the excellent reproducibility of NMR also makes it a particularly efficient profiling tool. On the other hand, the *non-destructive* and *non-invasive* nature of NMR allows the intact sample to be recovered after analysis, and makes *in vivo* measurements possible.

It should also be noted that NMR sample preparation is simple and that the analysis can be automated, in particular with the help of sample changers capable of analyzing up to a hundred NMR tubes in liquid-state spectroscopy. This automation allows a relatively low cost analysis, with an average cost of around €10 per sample (for 30 minutes of analysis, cost including depreciation and operation of the equipment). The purchase cost of the equipment, on the other hand, can be high for high-field spectrometers (around €1 M for a mid-range liquid-state NMR spectrometer, and up to €10 M for the highest magnetic fields).

The main limitation of NMR is still its sensitivity, due to the low polarization at room temperature (population difference between energy levels). It should be noted, however, that this sensitivity has been considerably improved since the introduction of NMR. This is due to two factors: (1) the increase in the intensity of the B_0 magentic field of superconducting magnets (the sensitivity evolves in $B_0^{2.5}$) and (2) the development of cryogenic probes whose electronic circuits are cooled in order to limit the thermal noise, allowing an increase in sensitivity by a factor of 3 to 4. Defining an NMR detection limit is not easy because it depends on the material used (magnetic field strength, quality factor of the probe used to detect the signal). In addition, the signal-to-noise ratio (SNR) increases with the square root of the number of repeated acquisitions on the sample studied. Therefore, there is no detection limit if an infinitely long analysis time is available. In practice, and in the case of complex samples, it is common to detect concentrations of around 1 $\mu mol.L^{-1}$ using ^1H NMR for small molecules in a mixture, with a mid-range NMR spectrometer.

Finally, it should be noted that NMR is a non-selective tool, that is, all of the molecules present in the sample give rise to an NMR signal from the moment that they are present in a sufficient quantity. This non-selectivity can both be seen as an advantage (NMR detects all that is detectable) and as a downfall. The signal from concentrated molecules can in fact mask that from weakly concentrated analytes, especially in the case of complex mixtures. In the case of complex samples, the large number of detectable analytes and the wide range of associated concentrations generates signal overlapping, which may affect the identification and quantification of analytes by NMR spectroscopy. In this case, the problem can be overcome by recording multi-dimensional spectra (two, three or more dimensions) in order to separate the NMR peaks according to several orthogonal dimensions. Multi-dimensional NMR spectroscopy, which appeared in 1971, is widely used for the structural elucidation of small molecules and macromolecules (proteins) in the liquid and solid phase, and more recently for the quantitative analysis of complex mixtures. On the other hand, there are NMR detection techniques used to filter the signals from specific classes of molecules, for example to remove the signals from macromolecules by favoring those from smaller molecules, or vice versa.

In summary, nuclear magnetic resonance spectroscopy can be used to accurately profile the major compounds of the large classes of molecules present in the egg matrix. The diversity of available techniques (relaxometry, spectroscopy, imaging) allows a very precise chemical and biochemical characterization of egg products, as illustrated in the following sections.

3.4.2. *Relaxation in NMR Spectroscopy*

3.4.2.1. *Principle*

NMR *relaxometry* is the measurement of *relaxation times*. This term designates the characteristic times at which the nuclear magnetization returns to its equilibrium state at the end of the NMR experiment. The longitudinal relaxation time (denoted T_1) characterizes the return to equilibrium of the nuclear magnetization parallel to the magnetic field B_0, whereas the transverse relaxation time (denoted T_2) governs the decay of the magnetization in the transverse plane (perpendicular to B_0). These parameters are physico-chemical quantities characteristic of the constituents of a given sample. They depend on many factors such as the temperature, the intensity of the magnetic field, and especially the physico-chemical environment of the nuclei. In particular, the measurement of relaxation times provides valuable information on molecular dynamics, as these parameters are intimately related to the mobility of molecules. Indeed, the relaxation times depend on the correlation time τ_c which represents the average time required for an object (a molecule for example) to perform a rotation of 1 radian. This dependence of the relaxation times on the mobility of the molecules makes it possible to probe their local environment, and thus to study the microstructure of complex samples in a non-destructive manner.

The relaxation times are measured using specific NMR experiments, which can be carried out using a relatively low magnetic field (typically < 1 T for applications in the food industry). This allows the use of low-cost devices (a few tens of thousands of euros) which are simple to use, but their spectral resolution is not sufficient to obtain exploitable spectra after Fourier transformation. The relaxation time measurements are therefore carried out directly in the time domain, that is, time-domain NMR (TD-NMR). In a complex sample, the detected signal represents the sum of the signals from the different components in the sample. The adjustment of this signal using a mathematical model therefore makes it possible to construct a distribution graph of the transverse or longitudinal relaxation times. It is also possible to obtain two-dimensional T_1-T_2 correlation maps. This approach is used in various fields ranging from the oil industry to the food industry. It should also be noted that these relaxation measurements may be combined with diffusometry measurements, in which the signal detected in NMR is linked to the molecular diffusion coefficient.

In the agri-food sector, these approaches are used for quality control applications, particularly to determine the water or lipid content in food. Relaxometry is also widely used to study the relationships between food properties and their microscopic structure. The reduced cost of relaxometers and their ease of use make time-domain NMR a key tool in this field.

The initial application of TD-NMR to the study of food involved the measurement of the solid/liquid ratio in foods from the measurement of transverse relaxation times (T_2). Magnetization in solids decreases much more rapidly than in liquids, and the FID is thus characterized by two main components: one corresponding to a short T_2 (rapid decay), associated with the solid part of the sample, and the other corresponding to a long T_2 (slow decay) associated with the liquid part. The measurement of these two components makes it possible to determine the percentage of solid fat, and is recognized as an official method in Europe and in the United States. Besides this leading application, time-domain NMR is used for a broad variety of agri-food applications [VAN 10]: for example, the study of the dynamics of water in cereals, the study of the impact of freezing on fruit and vegetables or the quality control of fish and meat.

3.4.2.2. *Application to the study of egg products*

Despite the widespread nature of NMR relaxometry in the analysis of agri-food products, applications for the study of egg products in the literature are still very limited. A whole egg study was conducted by Jayasundar *et al.* [JAY 97]. In this study, measurements were made *in vivo* under a relatively high magnetic field (4.7 T) in an MRI machine, and measurements were also performed using imaging. The relaxation time measurements were performed without signal localization. The detected signal thus comes from the whole egg, and the authors were able to separate, on the basis of the chemical shift, the relaxation times of egg white and yolk, respectively. In this context, the authors attempted to study the effect of egg cooking on relaxation times, in order to derive information on the modification of their microstructure. No significant variation in longitudinal relaxation times (T_1) was observed, but the authors did observe a significant variation in T_2 values during egg cooking. In particular, the transverse relaxation of egg yolk lipids is strongly modified, which was interpreted as an increase in lipid mobility due to a thermal denaturation process during the cooking of the egg.

In addition to this whole egg study, most relaxometry studies were performed *in vitro* on liquid egg fractions (egg white or yolk). Laghi *et al.* observed a significant effect of egg aging on longitudinal relaxation times in egg whites, which they attributed to a change in albumin mobility due to an increase in pH induced by loss of carbon dioxide [LAG 05]. Hills *et al.* also showed how the recording of two-dimensional T_1-T_2 maps makes it possible to characterize the denaturation and aggregation phenomena of proteins during the cooking of the egg [HIL 04]. Finally, Au *et al.* developed a method to study, by means of NMR relaxometry, the impact of the freeze–thaw process on the mobility of the main constituents of the egg [AU 16].

NMR relaxometry therefore has important prospects in the field of quality control of egg products; a change in distribution of relaxation times can be an effective marker of their deterioration.

3.4.3. *High-field NMR spectroscopy*

3.4.3.1. *Introduction*

High-field NMR spectroscopy (often referred to by the simple acronym NMR) has established itself as the go-to analytical technique in chemistry for the structural characterization of organic molecules, as well as in structural biology where it is one of the two techniques (alongside X-ray diffraction) for studying the structure and dynamics of proteins. Liquid-state NMR is also widely used, with mass spectrometry, in the field of metabolomics where it becomes possible to acquire a metabolic fingerprint of samples of biological relevance (extracts, biofluids etc.). Solid-state NMR is widely used in the study of materials and biomaterials. As a consequence, most chemical research laboratories (and some biology laboratories) are equipped with one or more NMR spectrometers. NMR is also very common in the industrial sector, particularly in the pharmaceutical field.

The development of NMR has been accompanied by a constant increase in the magnetic field and the associated measurement probes, thus making it possible to maximize the performance in terms of resolution and sensitivity (see section 3.1). This improvement in performance is reflected in the high cost, which makes accessing NMR equipment difficult for modest-sized companies or laboratories. Fortunately, most research labs offer analysis in the form of service or collaboration, and the most powerful equipment is organized in the form of national platforms open to external users. It is therefore possible for any laboratory or company to perform NMR analyses while benefiting from the experience of qualified personnel.

High-field NMR spectroscopy relies on the use of superconducting magnets. These are coils made of metal alloy (for example Nb-Ti) in which the current flows without energy dissipation by Joule effect, delivering a very intense, very stable, very homogeneous magnetic field B_0 in the center of the magnet (where the sample is placed). This superconductivity requires that the coil be placed at a very low temperature; it is thus permanently immersed in a liquid helium reservoir at –269 °C, itself surrounded by liquid nitrogen at –196 °C. Regular filling of liquid helium (several times a year) and of the liquid nitrogen tanks (weekly) must be carried out by qualified personnel.

For the study of samples in liquid phase, the samples are placed in a tube that is a few millimeters in diameter (usually 5 mm). The total volume of the sample analyzed is generally a few hundred microliters (600 to 800 µL for a tube that is 5 mm in diameter). The samples in powder form (solid, freeze-dried) are dissolved in a deuterated solvent; for samples that are already liquid, a small amount of deuterated compound is added. The use of a deuterated compound is necessary for a procedure called field-frequency lock (or simply "lock"). The lock consists of recording, in parallel with the NMR analysis, the signal of a deuterated compound in order to maintain the temporal stability of the magnetic field, which is essential for obtaining a high-quality spectrum. Moreover, since the resonance frequency of deuterium is different from that of the proton, this prevents the signal of the solvent from being present on the NMR spectrum. For liquid phase analyses, obtaining a high resolution is essential for the NMR spectrum to be exploitable. This requires a very homogeneous external magnetic field B_0 and the homogeneity of the field is ensured by additional coils (the "shims"), which generate small magnetic fields that the operator adjusts to compensate for the inhomogeneities of B_0.

The analysis of solid phase samples is complicated by the presence of numerous anisotropic interactions. These interactions are averaged at zero in the liquid samples due to

Brownian motions, but in the solid phase they become predominant and make the spectra difficult to exploit. The analysis of solid samples therefore requires time to place the sample in rapid rotation (several tens of kHz) at an inclined angle of 54.74° ("the magic angle") with respect to the magnetic field, in order to free some anisotropic interactions and to improve the spectral resolution. For this, the sample is placed in a small cylindrical container, called a rotor.

In both the liquid and solid phase, one of the great advantages of NMR spectroscopy is that many different experiments can be performed, corresponding to different ways of manipulating nuclear magnetizations. The simplest experiment allows the acquisition of a one-dimensional spectrum, but many sequences of radio-frequency pulses can be used to manipulate these magnetizations and obtain additional information. As indicated in section 3.1, multi-dimensional experiments are particularly useful for obtaining structural information because they make it possible to establish links between the different atoms constituting a molecule or a material. They are also useful for the quantitative analysis of mixtures, as they allow effective signal separation from different constituents [MAR 17].

In the agri-food industry, high-field NMR spectroscopy is widely used, particularly for the quality control of food products. For liquids, NMR is used for profiling purposes to authenticate the origin of beverages (wines, fruit juices, etc.), honey or edible oils [MAR 13]. Due to its quantitative nature and very high reproducibility, high-field NMR can also be used to measure changes in the natural abundance of isotopes such as ^{2}H or ^{13}C at each site of a molecule [JÉZ 17]. This isotopic NMR approach makes it possible to authenticate the origin of natural or synthetic molecules such as aromas (vanillin, caffeine, etc.). Finally, solid-state NMR allows the study of solid foods (cheese, meat, fish, etc.) by placing them in rotation at the magic angle of 54.74°.

3.4.3.2. *Application to the study of egg products*

To date, there are very few high-field NMR applications that study egg products. The vast majority of published works are studies aimed at characterizing the structure and dynamics of egg proteins (previously isolated and purified), such as lysozyme [RED 88]. A number of studies have also been performed on models to better understand the structure of vesicles formed by phosphatidylcholines (egg yolk lecithin) [FIN 71]. A quantitative ^{1}H NMR approach for the quality control of egg-based liquors has also been published [HOH 15]. However, high-field NMR is not routinely used as a technique for the characterization of egg products, despite its high potential for fine-grained analysis of their contents. Figure 3.18 illustrates this potential on samples (whole egg and polar fraction) analyzed using a high-field liquid-state spectrometer.

As shown in Figure 3.18a, high-field NMR shows significant differences between the spectra of a sterile and non-sterile whole egg. These preliminary results suggest that high-field NMR may be able to identify some degradation markers of egg products. Figure 3.18b shows how high-field NMR allows a very fine mapping of the molecular content of egg products, using a polar fraction as an example. However, the complexity of egg product samples leads to significantly overlapped signals in the 1D spectrum, which could hinder the targeted identification or quantification of some analytes. Figure 3.18c illustrates how some of these signals can be separated by two-dimensional NMR.

In summary, high-field NMR offers a very precise characterization of the molecular content of egg products, and could prove to be an accurate characterization technique to better understand degradation phenomena through the identification or quantification of degradation markers.

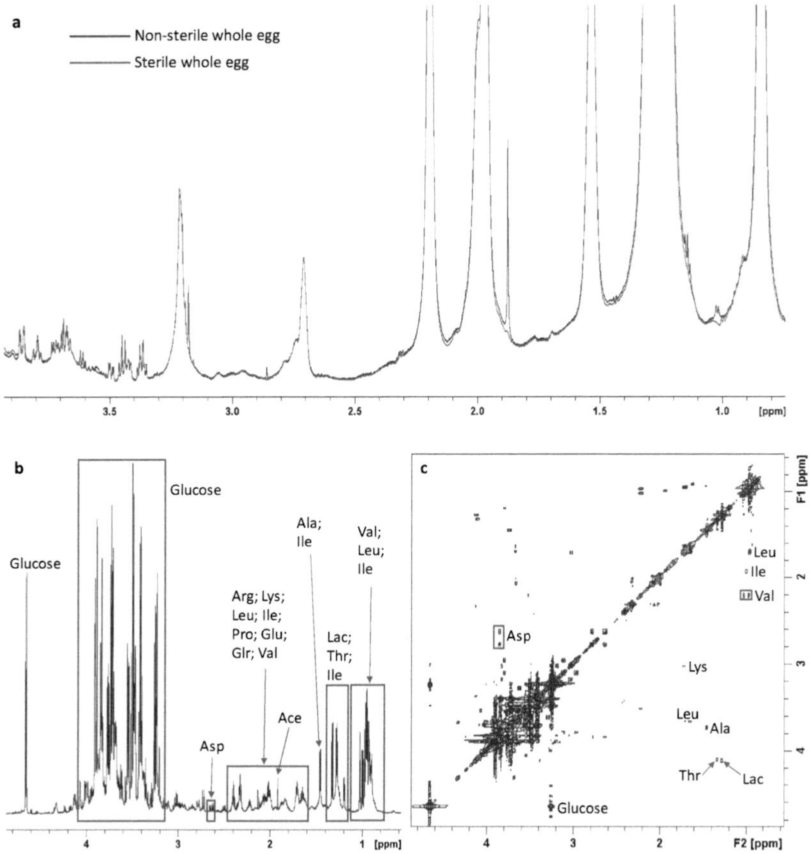

Figure 3.18. *Illustration of the potential of liquid-phase high-field NMR spectroscopy for the analysis of egg products. (a) The proton spectrum of a sterile egg sample compared to that of a non-sterile egg (freeze-dried samples dissolved in aqueous phase for analysis) reveals significant differences, which show that high-field NMR could identify potential degradation products. (b) The 1D spectrum of the polar fraction (lyophilized and then dissolved in an aqueous medium), recorded over 14 minutes with a solvent signal suppression method, allows the identification of the signals of a given number of metabolites. (c) A homonuclear 2D spectrum, recorded over 1h 30m on the same sample, allows the separation of signals of certain metabolites, overlapped on the 1D spectrum. The experiments were carried out on a 700 MHz (16.4 T) spectrometer at the Chemistry and Interdisciplinarity Laboratory: Synthesis, Analysis, Modeling, at the University of Nantes, France. Arg: Arginine; Lys: Lysine; Leu: Leucine; Ile: Isoleucine; Pro: Proline; Glu: Glutamate; Glr: Glutarate; Val: Valine; Ace: Acetate; Lac: Lactate; Thr: Threonine; Asp: Aspartate. For a color version of this figure, see www.iste.co.uk/goncalves/ovoproducts.zip*

3.5. Mass spectrometry

3.5.1. *Introduction*

As discussed in the previous chapters, acquisition of fingerprints can be done using different analytical techniques. Spectroscopic techniques are robust methods that allow access to rich intrinsic information. However, the analysis of trace metabolites may be a limiting factor of these techniques. In order to better characterize a sample to detect metabolites that mark a deterioration in food quality, mass spectrometry can provide valuable information on the molecular weight of metabolites [PIN 16]. In addition, this sensitive and specific technique allows access to metabolites, which are informative about the quality of the product. All mass spectrometers have three elements in common: an ion source, a mass analyzer and a detector (Figure 3.19). The molecules are introduced into the source where the ions are formed, according to different methods which will be explained later in this chapter. These are guided to the mass analyzer, where they are separated, and then to the detector. The spectra are recorded using a software that also allows control of the instrument. The analytes are then characterized by the measurement of mass-to-charge (m/z) ratios of whole ionized molecules (i.e. molecular ions) and of their fragmentation products (i.e. fragment ions), if there are any.

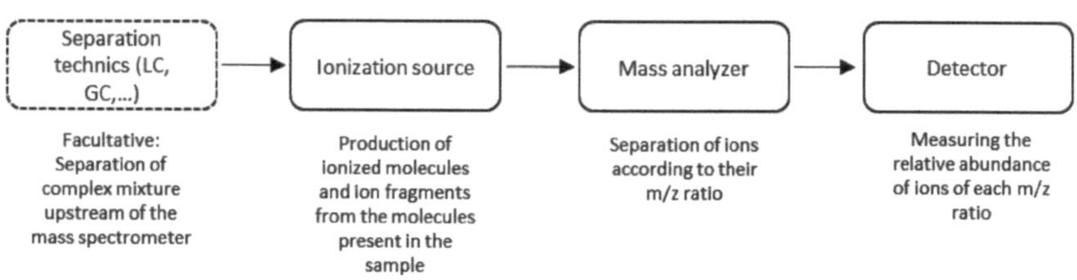

Figure 3.19. *Diagram of the different elements of a mass spectrometer*

The use of mass spectrometry requires prior purification of the samples. Given the diversity of metabolites present in food matrices, in terms of physical and chemical properties (polarity, volatility, stability (thermal, oxidative, hydrolytic)), exhaustive characterization is impossible. The target of the metabolism will be dependent on the protocols implemented for sample preparation (from the least selective solid–liquid extraction to a solid phase purification that can target a fraction of the metabolome, or even techniques that focus only on the headspace of the samples and therefore the most volatile compounds) [COU 14]. Choosing the sample preparation method is complex because it depends on (1) the chemical and physical properties of the compounds of interest, and (2) the food matrix, which is likely to interact with the analytes and cause degradation reactions, for example enzymatic reactions. Extraction conditions must be optimized to maximize analyte recovery [GAL 16].

Different acquisition modes, dependent on mass analyzers, are used in metabolomics. Each one has its own advantages and disadvantages; the choice of the mode mainly depends on the application and the level of selectivity sought [DUN 05]. Full scan acquisition mode remains a method of choice to obtain a global metabolomic fingerprint. In this mode, a mass range interval is defined and followed by the mass analyzer, for example from m/z 100 to m/z 1000. The sensitivity, the scan speed, the wide dynamic range and the high mass accuracy are the main parameters to consider in full scan acquisition mode. Indeed, the size of the metabolome investigated is unknown but is supposed to be very extensive with tens or even hundreds of compounds with the same molecular mass [WIS 13]. Thus, the ability of the mass analyzer to separate isobaric ions (i.e. ions with the same nominal mass), known as resolution, will be a particularly important asset [JUN 14]. The resolution R of a mass spectrometer is defined in a simplified way as the ratio $m/\Delta m$, where m is the mass of the ion considered and Δm the minimum difference between the peak considered and the nearest neighbor from which it can be distinguished. According to this definition, an instrument capable of distinguishing ions of m/z 100 and 100.1 has a resolution of 1000. The full scan mode is thus widely used with instruments with high resolution (TOF, time of flight, R = 5,000 to 10,000) or very high resolution (Orbitrap™, R = 10,000 to 100,000), or even ultra-high resolution (Fourier Transform Ion Cyclotron Resonance Mass Spectrometry (FT-ICR-MS, R up to 1,000,000). When the goal is to quantify given metabolites (targeted metabolomics), SIM (Single Ion Monitoring) or SRM (Selected Reaction Monitoring) modes may be preferred. In these modes, a list of m/z ratios to follow is defined and the mass analyzer will search specifically for these m/z in the sample. This acquisition technique does not diminish the importance of the resolving capacity of the instrument (especially in SIM mode and, to a lesser extent, in SRM mode).

With regard to the introduction of samples into the mass spectrometer, the direct introduction of the samples (by infusion or flow injection analysis) is commonly performed with ionization techniques at atmospheric pressure. With regard to what has been stated previously, this type of introduction requires mass analyzers at high or very high resolution [IBÁ 15]. The advantage of these techniques is their high throughput, especially when many samples need to be analyzed. However, it is very prone to matrix effects and especially to the phenomenon of ion suppression (competition for charges during the ionization process), which may limit its ability to cover the metabolome [ANT 05]. Thus, a separation technique upstream of the mass spectrometer is sometimes a prerequisite, even if it considerably lengthens the analysis time. The chromatographic separation upstream of the mass analyzer will make it possible to reduce matrix effects, since the various interfering compounds are distributed over the duration of the elution gradient. It thus makes it possible to separate isobaric compounds (same nominal mass) or isomers (same exact mass) which potentially have different physico-chemical properties and will therefore be eluted at different retention times [IBÁ 13]. These separation techniques, however, will also guide our ability to detect metabolites, as gas chromatography coupled with mass spectrometry (GC-MS) is more suitable for the analysis of semi-polar compounds than apolar and/or volatile compounds, while liquid chromatography coupled with mass spectrometry (LC-MS) is preferred for the analysis of polar and/or ionic compounds.

Given the investment cost associated with each piece of equipment, from €1,00,000 for a quadrupole mass analyzer to €5,00,000 or even €10,00,000 for ultra-high resolution or combined mass analyzers; it is very rare to have a large number of configurations in the same laboratory allowing a complete characterization of the metabolome. In addition, the degree of expertise associated with each piece of equipment can also be a limitation. Thus, data acquisition using fingerprints is often a matter of compromise, based on available equipment and research hypotheses.

3.5.2. *Gas chromatography coupled with mass spectrometry*

3.5.2.1. *An overview*

Developed in the 1960s, gas chromatography (GC) is the most widely used technique for analyzing non-thermolabile volatile or semi-volatile compounds. The mobile phase, called the carrier gas, is generally helium. The solutes are introduced either directly into the column when the sample is gaseous, or after volatilization of the liquid sample in the injection chamber, or by means of headspace techniques. The chromatographic separation depends principally on the volatilization temperatures of the solutes as well as their affinity with the stationary phase of the column, the nature of the mobile phase only affecting the chromatographic resolution. GC appears to be a technique of choice for analyzing apolar compounds such as fatty acids, volatile compounds or non-volatile polar compounds derived from primary metabolism, which are analyzed after derivatization. Indeed, to facilitate the analysis of polar, nonvolatile or thermolabile molecules, many chemical reactions have been developed to increase their volatility and thermal stability [PAS 08]. Two types of derivatizations are commonly encountered: oximation and silylation, the former being often followed by the latter.

Headspace techniques are widely used to extract volatile compounds from samples such as food. The term "headspace" (HS) refers to the gaseous phase (in equilibrium with the matrix or not) above a solid or liquid sample when it is placed in a sealed flask. There are two main types of HS sampling: static (S-HS) or dynamic (D-HS) headspace. For S-HS extraction, the sealed vial containing the sample is heated for a given period of time at a specified temperature, and the volatile compounds are distributed between the sample and the gas phase. A small fraction of the HS is taken and can be analyzed directly by GC. This technique is simple and robust, but relatively insensitive. To improve sensitivity, various devices based on different trapping materials and models have been developed, such as solid phase microextraction (SPME), which is widely used, or stir bar sorptive extraction (SBSE) [SNO 10, NOG 15]. Extraction by D-HS is based on the use of an inert gas stream for the continuous extraction of volatile compounds from a sample and their pre-concentration on an adsorbent trap, the most widely used being cryogenic. The volatile compounds are then released from the trap by heat and are transferred to the GC system. Whatever the HS technique, most of the volatile compounds are only partially recovered [SOR 15].

In most cases, gas chromatography is coupled with a mass spectrometer (GC-MS, short for gas chromatography–mass spectrometry). GC-MS has long been used in metabolomics because of its excellent separative capacity and robustness [CEV 12]. This robustness, especially when the ionization chosen is electroionization (EI), which is highly repeatable, allows the establishment and use of spectral databases that can be shared between laboratories or commercial databases, such as the National Institute of Standards and Technology (NIST) database.

Regarding the ionization of compounds, electroionization (EI) is the most commonly used in GC-MS. It makes it possible to obtain a characteristic and repeatable fragmentation of a molecule as stated above, and thus to determine the structure of compounds. However, it is a source of "hard" ionization, which causes a strong fragmentation and can sometimes be limiting. Indeed, if the compound is highly fragmented, then the molecular ion, the ion that corresponds to the whole unfragmented molecule, is no longer identifiable, and this can complicate the structural elucidation of the compounds. In addition, this high fragmentation leads to a redundancy of information (many fragment ions for the same retention time), which can prove troublesome during data processing. Another source of ionization used is positive or negative chemical ionization (PCI or NCI). This method is "softer" than electronic impact ionization, and so the molecular ion is often identifiable. However, the weak fragmentation induced by this type of source can limit the information required for the elucidation of the compound structure (the fragment ions making it possible to obtain valuable information on the chemical functions of a molecule). The latter, infrequently used for metabolomic analyses, is preferred for less stable compounds and/or in addition to an analysis with electroionization. Mass analyzers that are commonly coupled with gas chromatography are the quadrupole (unit resolution mass analyzer), time-of-flight and, more recently, the Orbitrap™ mass spectrometers [ANT 11].

3.5.2.2. *Examples of application in research into markers of food spoilage*

3.5.2.2.1. Targeted analysis

Targeted analysis by GC-MS is commonly applied to the study of the quality or origin of different food matrices. [CAV 07], for example, used this method to determine the origin of samples of extra-virgin olive oils. In this study, the profiling of five specific volatile biomarkers by targeted analysis effectively distinguished samples from different production sites and determined whether or not plantations were irrigated. This example shows the potential of such an approach for analyzing the quality of food matrices; however, its effectiveness is directly based on the relevance of the choice of markers analyzed. This type of profiling method has also been used to study the composition of volatile metabolites of various fruits and plants originating from Australia [SUL 15]. The objective was to evaluate the changes in quality occurring during the storage of these products in terms of production processes (including steps such as drying or grinding) or to determine the volatile compounds responsible for the difference in aroma among apple vinegars from different production processes [DAB 14].

These examples relate to the analysis of volatile compounds of various food products which have undergone spoilage during their production or storage. These compounds generally have a significant impact on the odor and taste of the products. Their analysis by gas chromatography is simpler than for compounds with higher boiling points; GC and GC-MS are analysis techniques that are both simple to apply and efficient. This approach (characterization of specific volatile compounds by GC-MS) has also been used for the study of various food matrices, such as dry ham [MAR 14] or edible oils [CHE 04].

Using GC-MS targeted assays for the specific study of spoilage markers, excluding volatile compounds, is very rare. One example is the work by Pizzoferrato *et al* [PIZ 93]. They used GC-MS for the targeted analysis of markers of physico-chemical degradation in egg products. They have developed a method for the characterization and quantification of sterols and their oxidative products in spray-dried egg products. In this example, the markers studied had already been identified as being involved in the degradation process of these products. Therefore, the method used was not intended to identify new potential degradation markers but to develop a technique for the simultaneous determination of several markers belonging to the same chemical family.

The lack of studies on the quality or spoilage of LEs, based on a GC-MS targeted metabolomic approach, can be partly explained by the complexity of the sample preparation steps. Indeed, the LE is composed of many elements, including abundant macromolecules, such as proteins, which make their analysis by GC difficult. Deproteinization steps for the samples are therefore necessary and must be completed by different separation or extraction steps in order to maximize the potential and simplicity of the analysis.

3.5.2.2.2. Non-targeted profiling

To date, few applications have been carried out on egg products. The non-targeted approach, however, has a very strong potential for detecting the activity of spoiling microorganisms. Metabolomics using MS can thus help establish early detection methods for these spoiling microorganisms by measuring changes in the profile of metabolites resulting from their activity in the food. In this sense, one of the most common strategies involves studying the volatile fraction of food in order to detect volatile organic compounds (VOCs) related to the growth of these microorganisms. VOCs can be analyzed in food simply by using fast extraction techniques of headspace samples. Food matrices can be compliant food, spoiled food or food matrices deliberately inoculated with strains of microorganisms likely to be responsible for spoilage. After a differential study of spoiled samples and compliant samples, multivariate analysis (by principal component analysis, PCA or partial least squares regression, PLS) can be performed to identify volatile compounds related to the microbiological spoilage of the product.

The feasibility of a VOC metabolic profiling method has thus been investigated to characterize the deterioration of various foods. For example, the spoilage of pork with *Salmonella typhimurium* has been characterized using GC-MS. After deconvolution of the data and multivariate analysis by PCA, a grouping of both naturally altered matrices and those altered by *Salmonella* were obtained. Univariate analyses subsequently revealed 16 volatile metabolites that mark spoilage by this bacterium [XU 10]. Similarly, an approach using differential metabolomic fingerprinting was used to identify marker metabolites of onions infected by neck rot, caused by the fungus *Botrytis allii*, and sour skin, caused by *Burkholderia cepacia*, two of the main diseases affecting onions during storage. Metabolic fingerprints obtained after extraction by HS-SPME of onion samples affected by these two pathogens and on control samples made it possible to highlight various volatile compounds which are markers of the two diseases, sometimes at concentrations of up to 100 times higher than those measured in onions from the control group [LI 11]. Similarly, a combination of three VOCs among the 27 identified by HS-SPME-GC-MS were proposed as spoilage indicators in raw chicken breast, namely ethanol, 3-methyl-butan-1-ol and acetic acid, attributed to different bacterial groups using partial least squares regression (PLS) [MIK 16].

"Electronic nose" systems, including those based on mass detection ("MS-based e-nose") can also be used to generate volatile profiles, constituting footprints that can be compared to those stored in dedicated GC-MS databases [WOJ 17]. For example, a study to identify egg yolk from seven different bird species based on their GC-analyzed fatty acid composition, volatile compounds, analyzed by HS-SPME-GC-MS, and electronic nose analysis data followed by multivariate analyses, identified specific markers that differentiated these samples from one another [WAN 14].

GC-MS is also a technique that is frequently used for profiling primary metabolites, that is, organic acids, sugars, amino acids, etc. [FIE 08]. Thus, other applications have been carried out using GC-MS which focused on the liquid polar fraction of samples after derivatization by methoximation followed by silylation. By performing metabolic profiling on this fraction, primary metabolites such as sugars and amino acids (dextrose, glycine, tyrosine and histidine) could be highlighted as potential markers of contamination of ground beef and poultry with *Escherichia coli* O157:H7 and *Salmonella spp* [CEV 12]. O'Gorman *et al.* [OGO 12] focused on polar (primary metabolites) and apolar (lipid and fatty acid) fractions of fungal metabolome in order to identify markers of post-harvest damage. Both fractions were analyzed using metabolic profiling, the first after derivatization by methoxymation and the second after transmethylation of free fatty acids and lipids. Following the application of multivariate statistical analyses (PCA and PLS) on the basis of the different metabolites identified in the profiles, the non-polar fraction appeared to be the most informative. It revealed a strong correlation between linoleic acid and pentadecanoic acid in damaged samples, indicating activation of a metabolic pathway when the fungus is damaged.

An example of GC-MS spectra corresponding to the analysis of a polar metabolic fraction of healthy and spoiled industrial LEs is presented in Figure 3.20.

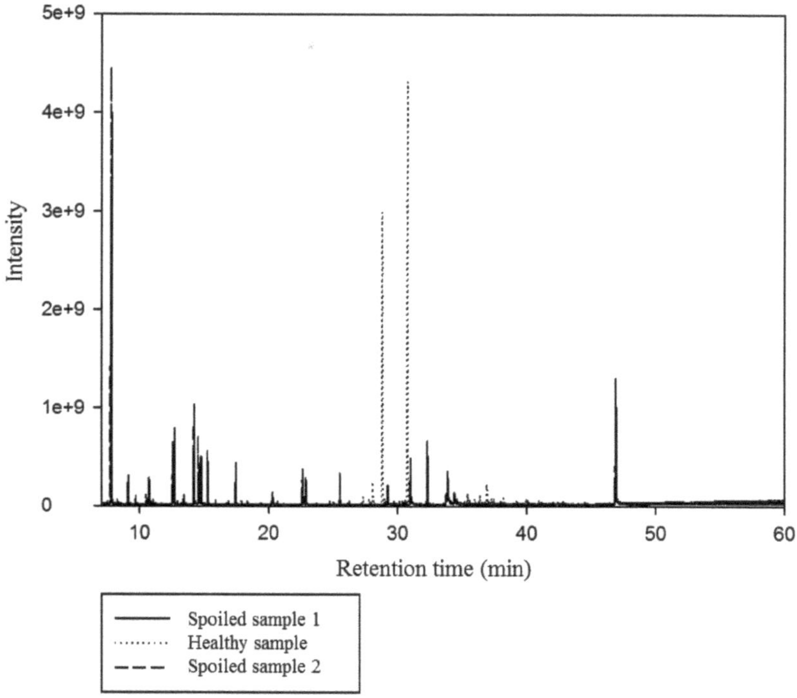

Figure 3.20. *TIC (Total Ion Chromatogram) GC of polar metabolic fractions of healthy and spoiled industrial LEs. The experiments were carried out in the Process Engineering for Environment and Food laboratory (GEPEA – UMR CNRS 6144) at the University of Nantes, France*

Although it is difficult to interpret a TIC at this scale, we can clearly distinguish specific peaks of some samples. If we focus on a smaller area of the spectrum (Figure 3.21), it allows us to see the chemical complexity of the analyzed fraction (in this case a polar fraction after cold ethanol deproteinization). It highlights certain differences in composition between the healthy sample and the two types of spoiled samples.

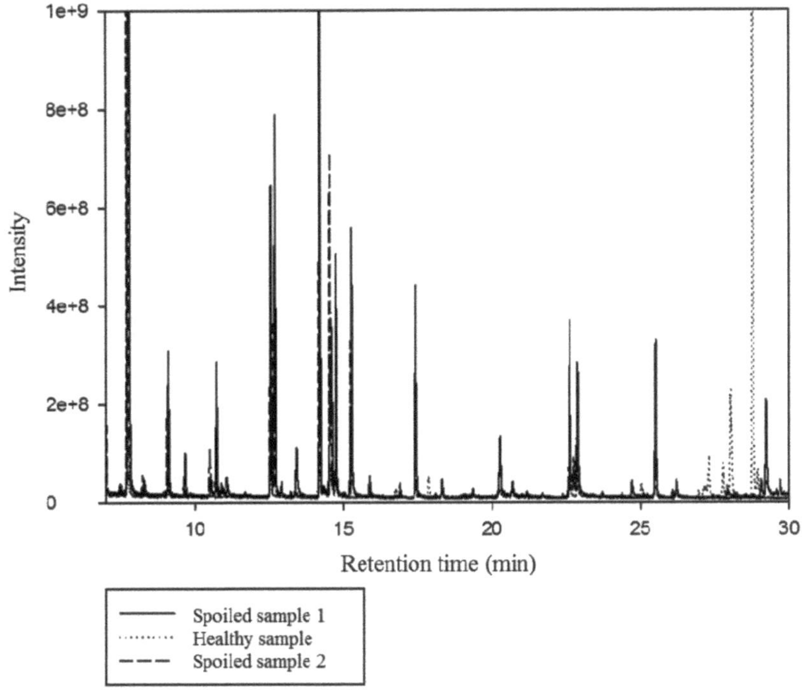

Figure 3.21. *TIC GC of polar metabolic fractions of healthy and spoiled industrial LEs – enlargement of a specific area. The experiments were carried out in the Process Engineering for Environment and Food laboratory (GEPEA – UMR CNRS 6144) at the University of Nantes, France*

The spectra obtained during the GC-MS analysis of food matrices, in particular LEs, are very complex due to the diversity of their chemical composition, often requiring the use of multivariate statistical approaches during their analysis. One of the most commonly used methods is principal component analysis. This method makes it possible to visualize the dispersion of the data and to potentially determine the discriminating capacity of a technique concerning the study of contrasting samples. The result of a principal component analysis performed on the data extracted from the previously presented spectra is presented in Figure 3.22.

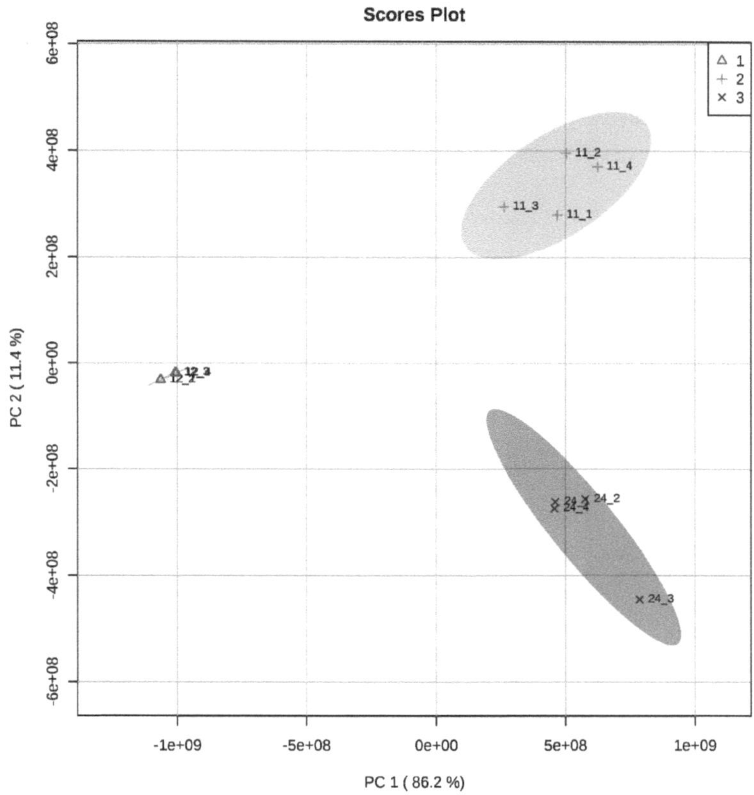

Figure 3.22. *Result of unsupervised principal component analysis performed on data from the GC-MS analysis of polar fractions from three industrial LE samples. The samples in red (12) correspond to healthy samples; the samples in green (11) and blue (24) correspond to spoiled industrial samples with two distinct types of spoilage. The experiments were carried out in the Process Engineering for Environment and Food laboratory (GEPEA – UMR CNRS 6144) at the University of Nantes, France. For a color version of this figure, see www.iste.co.uk/goncalves/ovoproducts.zip*

A principal component analysis makes it possible to very efficiently separate the samples. Axis 1 (x-axis), which represents 86.2% of the variance, corresponds to the difference between the healthy and spoiled samples. Axis 2 (y-axis), which represents 11.4% of the variance, is the difference between the two types of spoilage studied. This analysis highlights the discriminatory capacity of the non-targeted study of metabolic fractions by GC-MS for healthy and spoiled LE samples, and thus allows to see the very promising potential of non-targeted profiling techniques in the field of LE spoilage analysis.

3.5.2.2.3. *Toward a more overall characterization by GCxGC*

The combination of two or more separation steps in multidimensional configurations is used to improve the resolution of co-eluting metabolites in complex samples. The multidimensional separations, exploiting orthogonal phase chemistry of the chromatographic columns, make it possible to improve the number of analytes separated per unit of time (i.e. the peak capacity). In two-dimensional separations, all compounds eluting in the first dimension are then separated into a second dimension. GCxGC is the most-used combination (compared to LCxLC) [MUR 12]. Most GCxGC separations are based on the use of a first apolar column followed by a second polar column. GCxGC has become increasingly used in food analysis, mainly to study food composition. It has also been used to address food safety issues. For example, detection of microbiological spoilage of cucumber was investigated by non-targeted analysis of volatile metabolites using GCxGC-TOF-MS. Thanks to the separative power of two-dimensional chromatography, 214 volatile metabolites could be detected and described. An analysis of variance on these compounds revealed deterioration in concentration of 33 of these metabolites, including an increase in the concentrations of acetic, propanoic and butyric acids, several alcohols and a decrease in the concentration of furfural [JOH 11]. This technique was also used to characterize the deterioration of three different foods (bread, chicken egg and cucumber) using non-targeted metabolic profiling of the polar fraction of the samples after derivation by methoxymation and silylation. More than 2,000 chromatographic peaks were detected, among which were 13 metabolites (amino acids, biogenic amines and organic acids), were found to be markers of food spoilage. In eggs more particularly, 2.3-butanediol, putrescine, malonic and leucinic acids and nitrosotrimethylurea were identified as markers of food spoilage. It appeared that the deterioration processes were closely linked to certain metabolic pathways related to the bacteria that can colonize these foods during the spoilage processes [CHE 15].

3.5.3. *Liquid chromatography coupled with mass spectrometry*

3.5.3.1. *Overview of LC-MS*

Liquid chromatography (LC) coupled with mass spectrometry, introduced after GC-MS, has become a method of choice for metabolome analysis. The analytes are separated according to their solubility in the mobile phase and their affinity with the stationary phase. The main advantage of LC is to allow the analysis of compounds with low volatility and polar, thermolabile and/or high molecular weight, and which are incompatible with GC analysis. In general, this method does not require a derivation step before the injection. The separation of the compounds is related to the retention capacity of the molecules, that is to say, to their power of interaction with the stationary phase. The latter is the key element of a good chromatographic separation. The hydrophobic compounds will then be well separated with reversed-phase chromatography (C8, C18), while for the hydrophilic and neutral compounds, hydrophilic interaction chromatography (HILIC) will be the most suitable [THE 12]. There are several modes of ionization in LC-MS. The electrospray source ESI is widely used in metabolomics for the ionization of polar and non-volatile molecules. The main advantage of ESI is its "soft" ionization; this technique allows a very sensitive quantitative analysis. Nevertheless, the downfall of the ESI is its high sensitivity to matrix effects, in particular ion suppression phenomena, which can lead to a great variability in the measured signal. The ESI can be used in positive or negative ionization modes. The data obtained on the metabolome will be different depending on the mode chosen; some metabolites will be ionized more efficiently in one mode compared to the other. The ionization modes can therefore be complementary to provide additional information in order to obtain a comprehensive fingerprint [THE 12]. For the analysis of small polar molecules, atmospheric pressure chemical ionization (APCI) is generally used, and for small apolar molecules (inaccessible by ESI), ionization by photoionization (APPI) can be used [NOR 08]. Nevertheless, these two techniques are rarely mentioned in publications on metabolomic analyses. Mass analysers currently coupled with liquid chromatography are the quadrupole, ion trap (unit resolution mass analyzer), time of flight and Orbitrap™ mass spectrometers.

3.5.3.2. *Examples of application in search of food spoilage markers*

According to the studies reported in the literature, it is clear that generating global metabolic fingerprints using GC-MS is more effective than other techniques such as LC-MS in differentiating metabolic profiles related to the spoilage of food; this is mainly due to the fact that differences in chemical composition are found in volatile organic compounds or primary metabolites. Very few studies mention the use of LC-MS in a non-targeted approach to identify spoilage marker metabolites. [CEV 11] used this technique to identify bacterial marker metabolites in individual colonies of microorganisms (*Escherichia coli* O157:H7, *Salmonella typhimurium*, *Salmonella muenchen* and *Salmonella hartford*) or in a mixture. Only a differential peak was found on the metabolic profiles acquired by comparing all of the pathogens, demonstrating the limited power of the ESI to differentiate species of microorganisms based on the metabolites that are accessible using this analysis. However, if biogenic amines are more specifically targeted, LC is the method of choice. The production of biogenic amines is indeed associated with the activity of certain microorganisms in food. Biogenic amines thus serve as quality indicators for meat and meat products. Putrescine, cadaverine, histamine and tyramine are considered markers of freshness and can be used as indicators of microbial spoilage [HAL 94]. [HUT 13] thus quantified different biogenic amines in non-eviscerated pheasant meat using LC-MS by ESI in negative mode with a triple quadrupole (unit-resolution mass analyzer, SRM mode). Their goal was to determine a time and a temperature of storage which limits the formation of these compounds. On the basis of their results, authors have been able to recommend suitable T°/shelf life pairings adapted to the product. Although LC-MS applications are sparse for food spoilage, it has different applications in food science, whether for compositional studies [ROG 12], quality assessments, food safety [ZHA 12, TEN 13] or even the study of metabolic changes during food processing [ARA 12]. For example, the micro-oxygenation of wine, a treatment frequently used in its preparation, has been investigated using metabolic profiling. Indeed, the exposure of wines to excessive amounts of oxygen causes irreparable damage, leading to the production of unwanted flavors and possible bacterial damage. The effects of micro-oxygenation were thus studied through the non-targeted analysis of wine metabolites by LC-MS. Potential biomarkers of micro-oxygenation of wine have thus been highlighted. Of these, various pigments and tannins have been confirmed as already-known biomarkers. Other markers have also been highlighted, including caffeic acid, some organic acids and amino acids, which had never been previously proposed. Based on these results, new hypotheses concerning the formation and reactivity of wine pigments during micro-oxygenation could be proposed, revealing the contribution of non-targeted strategies to improving the understanding of wine chemistry [ARA 12].

3.5.3.3. *Toward greater use of LC-MS*

LC-MS, versatile as it is, does not allow the analysis of all metabolites and appears limited here for the identification of metabolite markers of food spoilage. Indeed, primary metabolite which often provide information on spoilage, are small polar compounds, eluted in the dead volume of reversed phase columns. Hydrophilic interaction chromatography (HILIC), suited for such compounds, will certainly allow LC to play an important role in this research, even if currently the robustness of these columns remains to be improved for metabolic profiling applications.

3.5.4. *Direct injection mass spectrometry*

3.5.4.1. *General overview on direct injection in mass spectrometry*

As previously mentioned, the advantage of the direct introduction of the samples is its high-throughput. An acquisition of data can be performed in a few seconds. Direct infusion (DI) and flow injection analysis (FIA) are the main techniques used to introduce liquid samples into the mass spectrometer. While infusion is the continuous ionization of a static sample, flow injection analysis involves injecting a sample into a continuous stream of mobile phase. Among the ionization techniques used at atmospheric pressure, ESI and, to a lesser extent, APCI are the techniques generally used in direct MS introduction [DRA 13]. Matrix-assisted laser desorption techniques for solid matrices can also be used, such as matrix-assisted laser desorption/ionization (MALDI); this technique is suitable for the analysis of peptides and proteins. The scope of this chapter is to cover low molecular weight metabolites; MALDI techniques will not be described. The recent introduction of desorption electrospray ionization (DESI) or direct analysis in real time (DART) has led to an increase in the number of applications that can be addressed by direct analysis in MS. These latter two techniques are considered as "ambient" ionization sources that require little or no sample preparation [IBÁ 15]. The sample is positioned at the entrance of a mass analyzer in an open environment and the desorbed molecules of the sample are ionized and transferred to the mass analyzer. These different ionization techniques are often or even always coupled with high-resolution mass analyzers (TOF, orbital trap, Fourier transform mass spectrometer). Indeed, without upstream chromatographic separation, the resolution of the isobaric ions requires the use of a mass analyzer with significant resolving power. The FT-ICR-MS mass spectrometer is able to analyze with a resolving power of up to 1,000,000 [ALL 12]. This ultra-high resolving power makes it possible to analyze extremely complex mixtures, such as food matrices. Mass accuracy below ppm (part per million) generally obtained with this type of instrument makes it possible to unambiguously determine a crude formula for a compound.

3.5.4.2. *What applications are there for these direct introduction techniques?*

Few studies to date have been conducted to search for food degradation markers using direct introduction techniques. Direct analysis of foods or food extracts using MS has, however, proven to be a relevant approach when rapid sample classification is required, such as in food authentication [GON 12, VAC 09, GAR 12] or rapid screening for fraud investigation [SAN 10]. This approach is also used to determine the origin of food [CAJ 11] or the composition of food [WU 04]. It could also be relevant to detecting degradation markers of egg products by using mass spectrometry in a reduced time.

3.6. Conclusion

The techniques used to assess the quality of LEs do not currently tackle the microbiological spoilage of these products in an industrial environment. The analytical approaches explored in this chapter are part of a more exhaustive study of these spoilage phenomena in the laboratory, aiming to better understand the metabolic complexity. Thus, the acquisition of non-targeted metabolic fingerprints, followed by a multivariate statistical study of the data, will make it possible to investigate these phenomena within LE matrices, with the aim of highlighting spoilage markers. Although many research projects have already demonstrated the feasibility of these approaches for the discovery of markers, their implementation in routine analyses is still limited. Whether the identification of these markers is carried out by LC-MS, GC-MS, direct analysis by NMR, or vibrational spectrometry (PIR, IRTF or Raman), they subsequently need to be validated for later use in routine analyses. This validation is still a topic of discussion in the scientific community and protocols need to be thought through and tested [RIE 15]. Once identified and validated, the markers could be followed by targeted approaches, which are easier to implement than non-targeted approaches. Moreover, targeted approaches use more common analyzers that are less expensive for laboratories in charge of control. The next chapter proposes to review the analytical possibilities that can use this information in routine analysis. Particular emphasis will be placed on the new generations of sensors that use the markers identified by the techniques mentioned in this chapter to finally draw conclusions about benchtop devices which implement these same approaches; these are already commercially available and are therefore easily deployable with reasonable effort.

3.7. Bibliography

[ALL 12] ALLWOOD J.W., PARKER D., BECKMANN M. *et al.*, "Fourier transform ion cyclotron resonance mass spectrometry for plant metabolite profiling and metabolite identification", *Methods Mol. Biol. Clifton NJ*, vol. 860, pp. 157–176, 2012.

[ANT 05] ANTIGNAC J.-P., DE WASCH K., MONTEAU F. *et al.*, "The ion suppression phenomenon in liquid chromatography – mass spectrometry and its consequences in the field of residue analysis", *Anal. Chim. Acta*, vol. 529, nos 1–2, pp. 129–136, 2005.

[ANT 11] ANTIGNAC J. *et al.*, "Mass spectrometry-based metabolomics applied to the chemical safety of food", *Trends Anal. Chem.*, vol. 30, no. 2, pp. 292–301, 2011.

[ARA 12] ARAPITSAS P. *et al.*, "A metabolomic approach to the study of wine micro-oxygenation", *PLOS ONE*, vol. 7, no. 5, p. e37783, May 2012.

[ARO 06] AROCA R., *Surface-Enhanced Vibrational Spectroscopy*, John Wiley & Sons, Chichester, 2006.

[AU 16] AU C., WANG T., ACEVEDO N.C., "Development of a low resolution 1H NMR spectroscopic technique for the study of matrix mobility in fresh and freeze-thawed hen egg yolk", *Food Chem*, vol. 204, pp. 159–166, 2016.

[BEL 00] BELL S.E., BURNS D.T., DENNIS A.C. *et al.*, "Rapid analysis of ecstasy and related phenethylamines in seized tablets by Raman spectroscopy", *Analyst*, vol. 125, no. 3, pp. 541–544, 2000.

[BER 09] BERTHOMIEU C., HIENERWADEL R., "Fourier transform infrared (FTIR) spectroscopy", *Photosynth. Res.*, vol. 101, nos 2–3, pp. 157–170, September 2009.

[BLU 12] BLUM M.-M., JOHN H., "Historical perspective and modern applications of Attenuated Total Reflectance – Fourier Transform Infrared Spectroscopy (ATR-FTIR): Modern applications of ATR-FTIR", *Drug Test. Anal.*, vol. 4, nos 3–4, pp. 298–302, March 2012.

[BLU 15] BLUME K., DIETRICH K., LILIENTHAL S. *et al.*, "Exploring the relationship between protein secondary structures, temperature-dependent viscosities, and technological treatments in egg yolk and LDL by FTIR and rheology", *Food Chem.*, vol. 173, pp. 584–593, April 2015.

[BOU 03] BOUŘ P., KEIDERLING T.A., "Empirical modeling of the peptide amide I band IR intensity in water solution", *J. Chem. Phys.*, vol. 119, no. 21, pp. 11253–11262, December 2003.

[CAB 05] CABONI M. *et al.*, "Effect of processing and storage on the chemical quality markers of spray-dried whole egg", *Food Chem.*, vol. 92, no. 2, pp. 293–303, September 2005.

[CAJ 11] CAJKA T., RIDDELLOVA K., TOMANIOVA M. *et al.*, "Ambient mass spectrometry employing a DART ion source for metabolomic fingerprinting/profiling: a powerful tool for beer origin recognition", *Metabolomics*, vol. 7, no. 4, pp. 500–508, 2011.

[CAN 02] CANET D., BOUBEL J-C., CANET SOULAS E., *La RMN : concepts, méthodes et applications*, Dunod, Paris, 2002.

[CAV 07] CAVALIERE B. *et al.*, "A metabolomic approach to the evaluation of the origin of extra virgin olive oil: a convenient statistical treatment of mass spectrometric analytical data", *J. Agric. Food Chem.*, vol. 55, no. 4, pp. 1454–1462, February 2007.

[CEV 11] CEVALLOS-CEVALLOS J.M., DANYLUK M.D., REYES-DE-CORCUERA J.I., "GC-MS based metabolomics for rapid simultaneous detection of Escherichia coli O157:H7, Salmonella Typhimurium, Salmonella Muenchen, and Salmonella Hartford in ground beef and chicken", *J. Food Sci.*, vol. 76, no. 4, pp. M238–M246, May 2011.

[CEV 12] CEVALLOS-CEVALLOS J.M., REYES-DE-CORCUERA J.I., "Metabolomics in food science", *Adv. Food Nutr. Res.*, vol. 67, pp. 1–24, 2012.

[CHA 84] CHARALAMBOUS G., *Analysis of Foods and Beverages: Modern Techniques*, Academic Press, Orlando, 1984.

[CHE 82] CHEN T.T., OWEN J.F., CHANG R.K. *et al.*, "Surface-enhanced raman-scattering of water adsorbed on silver electrodes", *Chem. Phys. Lett.*, vol. 89, no. 4, pp. 356–361, 1982.

[CHE 04] CHEMAT F., "Deterioration of edible oils during food processing by ultrasound", *Ultrason. Sonochem.*, vol. 11, no. 1, pp. 13–15, January 2004.

[CHE 15] CHENG J. *et al.*, "Evaluating potential markers of spoilage foods using a metabolic profiling approach", *Food Anal. Methods*, vol. 8, no. 5, pp. 1141–1149, May 2015.

[CHU 11] CHURCHILL M.E.A., SIBHATU H.M., UHLSON C.L., "Defining the structure and function of acyl-homoserine lactone autoinducers", in RUMBAUGH K.P. (ed.), *Quorum Sensing*, vol. 692, Humana Press, Totowa, 2011.

[COU 14] COURANT F., ANTIGNAC J.-P., DERVILLY-PINEL G. *et al.*, "Basics of mass spectrometry based metabolomics", *PROTEOMICS*, vol. 14, nos 21–22, pp. 2369–2388, 2014.

[DAB 14] DABIJA A., HATNEAN C.A., "Study concerning the quality of apple vinegar obtained through classical method", *J. Agroaliment. Process. Technol.*, vol. 20, no. 4, pp. 304–310, 2014.

[DAL 06] DALLE ZOTTE A., BERZAGHI P., JANSSON L.-M. *et al.*, "The use of near-infrared reflectance spectroscopy (NIRS) in the prediction of chemical composition of freeze-dried egg yolk and discrimination between different n–3 PUFA feeding sources", *Anim. Feed Sci. Technol.*, vol. 128, nos 1–2, pp. 108–121, May 2006.

[DEL 07] DELAROZADELGADO B. *et al.*, "Application of near-infrared microscopy (NIRM) for the detection of meat and bone meals in animal feeds: a tool for food and feed safety", *Food Chem.*, vol. 105, no. 3, pp. 1164–1170, 2007.

[DIE 93] DIEM M., *Introduction to Vibrational Spectroscopy*, John Wiley & Sons, New York, 1993.

[DRA 13] DRAPER J., LLOYD A.J., GOODACRE R. *et al.*, "Flow infusion electrospray ionisation mass spectrometry for high throughput, non-targeted metabolite fingerprinting: a review", *Metabolomics*, vol. 9, no. 1, pp. 4–29, 2013.

[DUN 05] DUNN W., ELLIS D., "Metabolomics: current analytical platforms and methodologies", *Trends Anal. Chem.*, vol. 24, no. 4, pp. 285–294, 2005.

[ERB 07] ERBERSDOBLER H.F., SOMOZA V., "Forty years of furosine – forty years of using Maillard reaction products as indicators of the nutritional quality of foods", *Mol. Nutr. Food Res.*, vol. 51, no. 4, pp. 423–430, April 2007.

[ERN 87] ERNST R.R., BODENHAUSEN G., WOKAUN A., *Principles of Nuclear Magnetic Resonance in One and Two Dimensions*, Oxford Science Publications, Oxford, 1987.

[EUR 04] EUROPEAN PARLIAMENT, COUNCIL OF THE EUROPEAN UNION, Regulation (EC) no. 853/2004 of the European Parliament and of the Council of 29 April 2004 laying down specific hygiene rules for food of animal origin, 2004.

[EUR 05] EUROPEAN COMMISSION, Commission Regulation (EC) no. 2073/2005 of 15 November 2005 on microbiological criteria for foodstuffs, 2005.

[FER 12] FERNÁNDEZ PIERNA J.A.P., VERMEULEN P., AMAND O. *et al.*, "NIR hyperspectral imaging spectroscopy and chemometrics for the detection of undesirable substances in food and feed", *Chemom. Intell. Lab. Syst.*, vol. 117, pp. 233–239, August 2012.

[FIE 02] FIEHN O., "Metabolomics – the link between genotypes and phenotypes", in TOWN C. (ed.), *Functional Genomics*, Springer, 2002.

[FIE 08] FIEHN O., "Extending the breadth of metabolite profiling by gas chromatography coupled to mass spectrometry", *Trends Anal. Chem.*, vol. 27, no. 3, pp. 261–269, 2008.

[FIN 71] FINER E.G., FLOOK A.G., HAUSER H., "The nature and origin of the NMR spectrum of unsonicated and sonicated aqueous egg yolk lecithin dispersions", *Biochim Biophys Acta*, vol. 260, pp. 59–69, 1971.

[FRI 96] FRIEDMAN M., "Nutritional value of proteins from different food sources. a review", *J. Agric. Food Chem.*, vol. 44, no. 1, pp. 6–29, January 1996.

[FUR 07] FURLAN P.-Y., SCOTT S.A., PEASLEE M.H., "FTIR-ATR study of pH effects on egg albumin secondary structure", *spectrosc. Lett.*, vol. 40, no. 3, pp. 475–482, May 2007.

[GAL 01] GALOBART J., BARROETA A.C., BAUCELLS M.D. *et al.*, "α-Tocopherol transfer efficiency and lipid oxidation in fresh and spray-dried eggs enriched with ω3-polyunsaturated fatty acids", *Poult. Sci.*, vol. 80, no. 10, pp. 1496–1505, 2001.

[GAL 16] GALLO M., FERRANTI P., "The evolution of analytical chemistry methods in foodomics", *J. Chromatogr. A*, vol. 1428, pp. 3–15, January 2016.

[GAR 12] GARRETT R., VAZ B.G., HOVELL A.M.C. *et al.*, "Arabica and robusta coffees: identification of major polar compounds and quantification of blends by direct-infusion electrospray ionization-mass spectrometry", *J. Agric. Food Chem.*, vol. 60, no. 17, pp. 4253–4258, May 2012.

[GIU 08] GIUNCHI A., BERARDINELLI A., RAGNI L. *et al.*, "Non-destructive freshness assessment of shell eggs using FT-NIR spectroscopy", *J. Food Eng.*, vol. 89, no. 2, pp. 142–148, November 2008.

[GON 12] GONZALEZ-DOMINGUEZ R., GARCIA-BARRERA T., GOMEZ-ARIZA J.-L., "Iberian ham typification by direct infusion electrospray and photospray ionization mass spectrometry fingerprinting", *Rapid Commun. Mass Spectrom.*, vol. 26, no. 7, pp. 835–844, April 2012.

[GOU 64] GOULDEN J.D.S., "Analysis of milk by infra-red absorption", *J. Dairy Res.*, vol. 31, no. 3, pp. 273–284, 1964.

[GRE 14] GREWAL M.K., JAISWAL P., JHA S.N., "Detection of poultry meat specific bacteria using FTIR spectroscopy and chemometrics", *J. Food Sci. Technol.*, July 2014.

[GUA 94] GUARDIOLA F., CODONY R., RAFECAS M. *et al.*, "Selective gas chromatographic determination of cholesterol in eggs", *J. Am. Oil Chem. Soc.*, vol. 71, no. 8, pp. 867–871, 1994.

[HAL 94] HALÁSZ A., BARÁTH Á., SIMON-SARKADI L. *et al.*, "Biogenic amines and their production by microorganisms in food", *Trends Food Sci. Technol.*, vol. 5, no. 2, pp. 42–49, 1994.

[HAN 11] HANSEN S., PEDERSEN-BJERGAARD S., RASMUSSEN K., "IR Spectrophotometry", *Introduction to Pharmaceutical Chemical Analysis*, John Wiley & Sons, 2011.

[HEA 76] HEANEY R.K., CURTIS R.F., "Some further studies on incubator reject eggs", *J. Sci. Food Agric.*, vol. 27, no. 11, pp. 1057–1063, 1976.

[HID 06] HIDALGO A., ROSSI M., POMPEI C., "Estimation of equivalent egg age through furosine analysis", *Food Chem.*, vol. 94, no. 4, pp. 608–612, March 2006.

[HID 08] HIDALGO A., FRANZETTI L., ROSSI M. *et al.*, "Chemical markers for the evaluation of raw material hygienic quality in egg products", *J. Agric. Food Chem.*, vol. 56, no. 4, pp. 1289–1297, February 2008.

[HID 95] HIDALGO A., ROSSI M., POMPEI C., "Furosine as a freshness parameter of shell eggs", *J. Agric. Food Chem.*, vol. 43, no. 6, pp. 1673–1677, 1995.

[HIL 04] HILLS B., BENAMIRA S., MARIGHETO N. *et al.*, "T1-T2 correlation analysis of complex foods", *Appl. Magn. Reson.*, vol. 26, pp. 543–560, 2004.

[HOH 15] HOHMANN M. *et al.*, "Quantitative 1H NMR analysis of egg yolk, alcohol, and total sugar content in egg liqueurs", *J. Agric. Food Chem.*, vol. 63, pp. 4112–4119, 2015.

[HUT 13] HUTAŘOVÁ Z., VEČEREK V., STEINHAUSEROVÁ I. *et al.*, "Effects of storage temperature on biogenic amine concentrations in meat of uneviscerated pheasants (Phasianus colchicus)", *Acta Vet. Brno*, no. 82, pp. 61–65, 2013.

[IBÁ 13] IBÁÑEZ C., GARCÍA-CAÑAS V., VALDÉS A. *et al.*, "Novel MS-based approaches and applications in food metabolomics", *Trends Anal. Chem.*, vol. 52, pp. 100–111, 2013.

[IBÁ 15] IBÁÑEZ C., SIMO C., GARCÍA-CAÑAS V. *et al.*, "The role of direct high-resolution mass spectrometry in foodomics", *Anal. Bioanal. Chem.*, vol. 407, no. 21, pp. 6275–6287, August 2015.

[JAY 97] JAYASUNDAR R., AYYAR S., RAGHUNATHAN P., "Proton resonance imaging and relaxation in raw and cooked hen eggs", *Magn. Reson. Imaging*, vol. 15, pp. 709–717, 1997.

[JÉZ 17] JÉZÉQUEL T., JOUBERT V., GIRAUDEAU P. *et al.*, "The new face of isotopic NMR at natural abundance", *Magn. Reson. Chem.*, vol. 55, pp. 77–90, 2017.

[JOH 11] JOHANNINGSMEIER S.D., MCFEETERS R.F., "Detection of volatile spoilage metabolites in fermented cucumbers using nontargeted, comprehensive 2-dimensional gas chromatography-time-of-flight mass spectrometry (GCxGC-TOFMS)", *J. Food Sci.*, vol. 76, no. 1, pp. C168–C177, February 2011.

[JUN 14] JUNOT C., FENAILLE F., COLSCH B. *et al.*, "High resolution mass spectrometry based techniques at the crossroads of metabolic pathways", *Mass Spectrom. Rev.*, vol. 33, no. 6, pp. 471–500, December 2014.

[KAR 03] KAROUI R., MAZEROLLES G., DUFOUR É., "Spectroscopic techniques coupled with chemometric tools for structure and texture determinations in dairy products", *Int. Dairy J.*, vol. 13, no. 8, pp. 607–620, January 2003.

[KAR 06] KAROUI R., KEMPS B., BAMELIS F. *et al.*, "Methods to evaluate egg freshness in research and industry: a review", *Eur. Food Res. Technol.*, vol. 222, nos 5–6, pp. 727–732, March 2006.

[KAR 09] KAROUI R., KETELAERE B.D., KEMPS B. *et al.*, "Eggs and egg products", in SUN D.-W. (ed.), *Infrared Spectroscopy for Food Quality Analysis and Control*, Academic Press, San Diego, 2009.

[KEE 10] KEELER J., *Understanding NMR Spectroscopy,* 2nd ed., John Wiley & Sons, Cambridge, 2010.

[KNE 03] KNEIPP J., SCHUT T.B., KLIFFEN M. *et al.*, "Characterization of breast duct epithelia: a Raman spectroscopic study", *Vib. Spectrosc.*, vol. 32, no. 1, pp. 67–74, 2003.

[KOG 63] KOGELNIK H., PORTO S.P.S., "Continuous helium-neon red laser as a Raman source", *J. Opt. Soc. Am.*, vol. 53, no. 12, pp. 1446–1447, 1963.

[KOP 04] KOPKA J., FERNIE A., WECKWERTH W. *et al.*, "Metabolite profiling in plant biology: platforms and destinations", *Genome Biol.*, vol. 5, no. 6, p. 109, 2004.

[KRI 05] KRISHNA C.M., SOCKALINGUM G.D., KEGELAER G. *et al.*, "Micro-Raman spectroscopy of mixed cancer cell populations", *Vib. Spectrosc.*, vol. 38, nos 1–2, pp. 95–100, July 2005.

[KUD 05] KUDELSKI A., "Characterization of thiolate-based mono- and bilayers by vibrational spectroscopy: a review", *Vib. Spectrosc.*, vol. 39, no. 2, pp. 200–213, October 2005.

[LAG 05] LAGHI L., CREMONINI M.M., PLACUCCI G. *et al.*, "A proton NMR relaxation study of hen egg quality", *Magn. Reson. Imaging*, vol. 23, no. 3, pp. 501–510, April 2005.

[LAN 11] LANG I., HODAC L., FRIEDL T. *et al.*, "Fatty acid profiles and their distribution patterns in microalgae: a comprehensive analysis of more than 2000 strains from the SAG culture collection", *BMC Plant Biol.*, vol. 11, no. 1, p. 124, 2011.

[LEC 77] LECHEVALIER M.P., MOSS C.W., "Lipids in bacterial taxonomy – taxonomist's view", *CRC Crit. Rev. Microbiol.*, vol. 5, no. 2, pp. 109–210, 1977.

[LIC 02] LI-CHAN E.C.Y., ISMAIL A.A., SEDMAN J. *et al.*, "Vibrational spectroscopy of food and food products", *Handb. Vib. Spectrosc.*, 2002.

[LI 11] LI C., SCHMIDT N.E., GITAITIS R., "Detection of onion postharvest diseases by analyses of headspace volatiles using a gas sensor array and GC-MS", *LWT – Food Sci. Technol.*, vol. 44, no. 4, pp. 1019–1025, 2011.

[LIN 11] LIN H., ZHAO J., SUN L. *et al.*, "Freshness measurement of eggs using near infrared (NIR) spectroscopy and multivariate data analysis", *Innov. Food Sci. Emerg. Technol.*, vol. 12, no. 2, pp. 182–186, April 2011.

[LIT 82] LITTMANN S., ACKER L., SCHULTE E., "Beurteilung des hygienischen Zustandes von Eiprodukten vor der Pasteurisierung über das Muster der organischen Säuren", *Z. Für Lebensm.-Unters. Forsch.*, vol. 175, no. 2, pp. 106–112, March 1982.

[LUC 96] LUCISANO M., HIDALGO A., COMELLI E.M. *et al.*, "Evolution of chemical and physical albumen characteristics during the storage of shell eggs", *J. Agric. Food Chem.*, vol. 44, no. 5, pp. 1235–1240, 1996.

[MAQ 02] MAQUELIN K. *et al.*, "Identification of medically relevant microorganisms by vibrational spectroscopy", *J. Microbiol. Methods*, vol. 51, no. 3, pp. 255–271, 2002.

[MAR 13] MARCONE M.F., WANG S., ALBABISH W. *et al.*, "Diverse food-based applications of nuclear magnetic resonance (NMR) technology", *Food Res. Int.*, vol. 51, no. 2, pp. 729–747, 2013.

[MAR 14] MARUŠIĆ N., VIDAČEK S., JANČI T. *et al.*, "Determination of volatile compounds and quality parameters of traditional Istrian dry-cured ham", *Meat Sci.*, vol. 96, no. 4, pp. 1409–1416, April 2014.

[MAR 17] MARCHAND J., MARTINEAU E., GUITTON Y. *et al.*, "Multidimensional NMR approaches towards highly resolved, sensitive and high-throughput quantitative metabolomics", *Curr. Op. Biotechnol.*, vol. 43, pp. 49–55, 2017.

[MIE 08] MIE G., "Contributions to the optics of turbid media, particularly of colloidal metal solutions", *Ann. Phys.*, vol. 330, no. 3, pp. 377–445, 1908.

[MIK 16] MIKŠ-KRAJNIK M., YOON Y-J., UKUKU D.O. *et al.*, "Identification and quantification of volatile chemical spoilage indexes associated with bacterial growth dynamics in aerobically stored chicken", *J. Food Sci.*, vol. 81, no. 8, pp. M2006–M2014, August 2016.

[MIN 17] Ministère de l'agriculture, de l'agroalimentaire et de la forêt, Vademecum sectoriel production d'ovoproduits et d'oeufs liquides – Version 2.0, February 2017.

[MOR 89] MORRIS C.E., HOERNING E.F., ANGELO A.J.S., "Uracil as a potentially useful indicator of spoilage in egg products", *J. Food Sci.*, vol. 54, no. 3, pp. 581–583, May 1989.

[MUR 82] MURPHY D.V., VONRABEN K.U., CHANG R.K. *et al.*, "Surface-enhanced hyper-Raman scattering from So32- adsorbed on Ag powder", *Chem. Phys. Lett.*, vol. 85, no. 1, pp. 43–47, 1982.

[MUR 12] MURRAY J.A., "Qualitative and quantitative approaches in comprehensive two-dimensional gas chromatography", *J. Chromatogr. A*, vol. 1261, pp. 58–68, 2012.

[NAU 10] NAU F., GUÉRIN-DUBIARD C., BARON F. *et al.*, *Science et technologie de l'œuf, volume 2 : de l'œuf aux ovoproduits*, Lavoisier, 2010.

[NAW 13] NAWROCKA A., LAMORSK J., "Determination of food quality by using spectroscopic methods", in GRUNDAS S. (ed.), *Advances in Agrophysical Research*, InTech, 2013.

[NOG 15] NOGUEIRA J.M.F., "Stir-bar sorptive extraction: 15 years making sample preparation more environment-friendly", *Trends Anal. Chem.*, vol. 71, pp. 214–223, September 2015.

[NOR 61] NORRIS K.H., BUTLER W.L., "Techniques for obtaining absorption spectra on intact biological samples", *IRE Trans. Bio-Med. Electron.*, vol. 8, no. 3, pp. 153–157, 1961.

[NOR 08] NORDSTRÖM A., WANT E., NORTHEN T. *et al.*, "Multiple ionisation mass spectrometry strategy used to reveal the complexity of metabolomics", *Anal. Chem.*, vol. 80, pp. 421–429, 2008.

[NOT 04] NOTINGHER I. *et al.*, "Discrimination between ricin and sulphur mustard toxicity in vitro using Raman spectroscopy", *J. R. Soc. Interface*, vol. 1, no. 1, pp. 79–90, 2004.

[OGO 12] O'GORMAN A., BARRY-RYAN C., FRIAS J.M., "Evaluation and identification of markers of damage in mushrooms (Agaricus bisporus) postharvest using a GC/MS metabolic profiling approach", *Metabolomics*, vol. 8, no. 1, pp. 120–132, 2012.

[OLE 62] O'LEARY W.M., "The fatty acids of bacteria", *Bacteriol. Rev.*, vol. 26, no. 4, p. 421, 1962.

[OSB 93] OSBORNE B.G., FEARN T., HINDLE P.T. *et al.*, *Practical NIR Spectroscopy with Applications in Food and Beverage Analysis,* 2nd ed., Longman Scientific and Technical, Harlow, 1993.

[OSB 00] OSBORNE B.G., "Near-infrared spectroscopy in food analysis", in MEYERS R.A. (ed.), *Encyclopedia of Analytical Chemistry*, John Wiley & Sons, Chichester, 2000.

[OTT 91] OTTO C., HOEBEN F.P., GREVE J., "Surface-enhanced Raman-scattering of the complexes of silver with adenine and damp", *J. Raman Spectrosc.*, vol. 22, no. 12, pp. 791–796, December 1991.

[PAR 80] PARRY A.E., ROBINSON D.S., WEDZICHA B.L., "An enzymic assay for β-hydroxybutyric acid in liquid whole egg", *J. Sci. Food Agric.*, vol. 31, no. 9, pp. 905–910, 1980.

[PAS 08] PASIKANTI K., HO P., CHAN E., "Gas chromatography/mass spectrometry in metabolic profiling of biological fluids", *J. Chromatogr. B*, vol. 871, no. 2, pp. 202–211, 2008.

[PIC 06] PICINELLI LOBO A., SUÁREZ VALLES B., FERNÁNDEZ TASCÓN N. *et al.*, "Calibration models for routine analysis of cider by mid-infrared spectroscopy", *LWT – Food Sci. Technol.*, vol. 39, no. 9, pp. 1026–1032, November 2006.

[PIN 16] PINU F.R., "Early detection of food pathogens and food spoilage microorganisms: application of metabolomics", *Trends Food Sci. Technol.*, vol. 54, pp. 213–215, 2016.

[PIZ 93] PIZZOFERRATO L., NICOLI S., LINTAS C., "GC-MS characterization and quantification of sterols and cholesterol oxidation products", *Chromatographia*, vol. 35, no. 5, pp. 269–274, 1993.

[POW 86] POWELL J.R., WASACZ F.M., JAKOBSEN R.J., "An algorithm for the reproducible spectral subtraction of water from the FT-IR spectra of proteins in dilute solutions and adsorbed monolayers", *Appl. Spectrosc.*, vol. 40, no. 3, pp. 339–344, 1986.

[RAM 28] RAMAN C.V., KRISHNAN K.S., "A new type of secondary radiation", *Nature*, vol. 121, pp. 501–502, January 1928.

[RED 88] REDFIELD C., DOBSON C.M., "Sequential 1H NMR assignments and secondary structure of hen egg white lysozyme in solution", *Biochemistry (Mosc.)*, vol. 27, pp. 122–136, 1988.

[RIE 15] RIEDL J., ESSLINGER S., FAUHL-HASSEK C., "Review of validation and reporting of non-targeted fingerprinting approaches for food authentication", *Anal. Chim. Acta*, vol. 885, pp. 17–32, 2015.

[ROB 75] ROBINSON D.S., BARNES E.M., TAYLOR J., "Occurrence of β-hydroxybutyric acid in incubator reject eggs", *J. Sci. Food Agric.*, vol. 26, no. 1, pp. 91–98, 1975.

[ROG 12] ROGACHEV I., AHARONI A., "UPLC-MS-based metabolite analysis in tomato", in HARDY N.W., HALL R.D. (eds), *Plant Metabolomics: Methods and Protocols*, Humana Press, Totowa, 2012.

[ROM 06] ROMEO M., MOHLENHOFF B., DIEM M., "Infrared micro-spectroscopy of human cells: causes for the spectral variance of oral mucosa (buccal) cells", *Vib. Spectrosc.*, vol. 42, no. 1, pp. 9–14, October 2006.

[ROS 95] ROSSI M., POMPEI C., HIDALGO A., "Freshness criteria based on physical and chemical modifications occurring in eggs during aging", *Ital. J. Food Sci. Riv. Ital. Sci. Degli Alimenti*, 1995.

[ROS 13] ROSSI M. *et al.*, "Developments in understanding and assessment of egg and egg product quality over the last century", *Worlds Poult. Sci. J.*, vol. 69, no. 02, pp. 414–429, June 2013.

[RUL 15] RUL F., MONNET V., "How microbes communicate in food: a review of signaling molecules and their impact on food quality", *Curr. Opin. Food Sci.*, vol. 2, pp. 100–105, April 2015.

[RYD 99] RYDER A.G., O'CONNOR G.M., GLYNN T.J., "Identifications and quantitative measurements of narcotics in solid mixtures using near-IR Raman spectroscopy and multivariate analysis", *J. Forensic Sci.*, vol. 44, no. 5, pp. 1013–1019, 1999.

[SAL 72] SALWIN H., STARUSZKIEWICZ W.F.J., BOND J.F., "Formation of β-hydroxybutyric acid in eggs during incubation.", *J. Assoc. Off. Anal. Chem.*, vol. 55, no. 3, pp. 458–460, May 1972.

[SAL 04] SALMINEN S., VON WRIGHT A., *Lactic Acid Bacteria: Microbiological and Functional Aspects,* 3rd ed., CRC Press, Boca Raton, 2004.

[SAN 88] SANCHEZ-RUIZ J.M., MARTINEZ-CARRION M., "A Fourier transform infrared spectroscopic study of the phosphoserine residues in hen egg phosvitin and ovalbumin", *Biochemistry (Mosc.)*, vol. 27, no. 9, pp. 3338–3342, 1988.

[SAN 10] SANVIDO G.B. *et al.*, "Fast screening and secure confirmation of milk powder adulteration with maltodextrin via electrospray ionization–mass spectrometry [ESI(+)–MS] and selective enzymatic hydrolysis", *J. Agric. Food Chem.*, vol. 58, no. 17, pp. 9407–9412, 2010.

[SCH 97] SCHRADER B., MOORE D.S., "Nomenclature, symbols, units, and their usage in spectrochemical analysis XVIII. Laser-based molecular spectroscopy for chemical analysis – Raman scattering processes", *Pure Appl. Chem.*, vol. 69, no. 7, pp. 1451–1468, July 1997.

[SCH 00] SCHUSTER K.C., URLAUB E., GAPES J., "Single-cell analysis of bacteria by Raman microscopy: spectral information on the chemical composition of cells and on the heterogeneity in a culture", *J. Microbiol. Methods*, vol. 42, no. 1, pp. 29–38, 2000.

[SHA 06] SHANMUKH S., JONES L., DRISKELL J. *et al.*, "Rapid and sensitive detection of respiratory virus molecular signatures using a silver nanorod array SERS substrate", *Nano Lett.*, vol. 6, no. 11, pp. 2630–2636, 2006.

[SNO 10] SNOW N.H., BULLOCK G.P., "Novel techniques for enhancing sensitivity in static headspace extraction-gas chromatography", *J. Chromatogr. A*, vol. 1217, no. 16, pp. 2726–2735, April 2010.

[SOR 15] SORIA A.C., GARCÍA-SARRIÓ M.J., SANZ M.L., "Volatile sampling by headspace techniques", *Trends Anal. Chem.*, vol. 71, pp. 85–99, September 2015.

[SUL 15] SULTANBAWA Y., WILLIAMS D., CHALIHA M. *et al.*, *Changes in Quality and Bioactivity of Native Foods During Storage*, Rural Industries Research and Development Corporation (Australia), 2015.

[TEN 95] TENENHAUS C., GAUCHI M., MENARDO J.P., "Régression PLS et applications", *Rev. Stat. Appliquée*, vol. 43, no. 1, pp. 7–63, 1995.

[TEN 13] TENGSTRAND E., ROSÉN J., HELLENÄS K.-E. *et al.*, "A concept study on non-targeted screening for chemical contaminants in food using liquid chromatography–mass spectrometry in combination with a metabolomics approach", *Anal. Bioanal. Chem.*, vol. 405, no. 4, pp. 1237–1243, 2013.

[THE 12] THEODORIDIS G., GIKA H., WANT E. *et al.*, "Liquid chromatography–mass spectrometry based global metabolite profiling: a review", *Anal. Chim. Acta*, vol. 711, pp. 7–16, 2012.

[UNE 10] UNECE WORKING PARTY ON AGRICULTURAL QUALITY STANDARDS, UNECE standard egg-2 concerning the marketing and commercial quality control of egg products – 2010 edition, 2010.

[UTA 85] UTAGAWA T., MORISAWA H., YAMANAKA S. *et al.*, "Properties of Nucleoside Phosphorylase from *Enterobacter aerogenes*", *Agric. Biol. Chem.*, vol. 49, no. 11, pp. 3239–3246, November 1985.

[VAC 09] VACLAVIK L., CAJKA T., HRBEK V. *et al.*, "Ambient mass spectrometry employing direct analysis in real time (DART) ion source for olive oil quality and authenticity assessment", *Anal. Chim. Acta*, vol. 645, nos 1–2, pp. 56–63, 2009.

[VAN 92a] VAN DE VOORT F.R., "Fourier transform infrared spectroscopy applied to food analysis", *Food Res. Int.*, vol. 25, no. 5, pp. 397–403, 1992.

[VAN 92b] VAN DE VOORT F.R., SEDMAN J., ISMAIL A.A., "A Rapid FTIR quality control method for fat and moisture determination in butter", *Food Res. Int.*, vol. 25., no. 3 pp. 193–198, 1992.

[VAN 93] VAN DE VOORT F., SEDMAN J., ISMAIL A.A., "A rapid FTIR quality-control method for determining fat and moisture in high-fat products", *Food Chem.*, vol. 48, no. 2, pp. 213–221, 1993.

[VAN 02] VANKEIRSBILCK T. *et al.*, "Applications of Raman spectroscopy in pharmaceutical analysis", *Trends Anal. Chem.*, vol. 21, no. 12, pp. 869–877, 2002.

[VAN 10] VAN DUYNHOVEN J., VODA A., WITEK M. *et al.*, "Time-domain NMR applied to food products", *Annu. Rep. NMR Spectrosc.*, vol. 69, pp. 145–197, 2010.

[VOD 98] VO-DINH T., "Surface-enhanced Raman spectroscopy using metallic nanostructures", *Trends Anal. Chem.*, vol. 17, nos 8–9, pp. 557–582, September 1998.

[WAH 93] WAHLE K.W., HOPPE P.P., MCLNTOSHA G., "Effects of storage and various intrinsic vitamin E concentrations on lipid oxidation in dried egg powders", *J. Sci. Food Agric.*, vol. 61, no. 4, pp. 463–469, 1993.

[WAN 14] WANG Q. *et al.*, "Discriminating eggs from different poultry species by fatty acids and volatiles profiling: comparison of SPME-GC/MS, electronic nose, and principal component analysis method: fatty acid and volatile profiles of avian eggs", *Eur. J. Lipid Sci. Technol.*, vol. 116, no. 8, pp. 1044–1053, August 2014.

[WAR 15] WARD B., "Bacterial energy metabolism", *Molecular Medical Microbiology*, Elsevier, 2015.

[WEH 88] WEHLING R.L., PIERCE M.M., FRONING G.W., "Determination of moisture, fat and protein in spray-dried whole egg by near infrared reflectance spectroscopy", *J. Food Sci.*, vol. 53, no. 5, pp. 1355–1359, 1988.

[WIS 13] WISHART D., JEWISON T., GUO A.C. *et al.*, "HMDB 3.0 – the human metabolome database in 2013", *Nucleic Acids Res.*, vol. 41, pp. D801–D807, 2013.

[WOJ 17] WOJNOWSKI W., MAJCHRZAK T., DYMERSKI T. *et al.*, "Electronic noses: powerful tools in meat quality assessment", *Meat Sci.*, vol. 131, pp. 119–131, September 2017.

[WOL 89] WOLD S., KETTANEH-WOLD N., SKAGERBERG B., "Nonlinear PLS modeling", *Chemom. Intell. Lab. Syst.*, vol. 7, nos 1–2, pp. 53–65, 1989.

[WU 04] WU Z., RODGERS R.P., MARSHALL A.G., "Characterization of vegetable oils: detailed compositional fingerprints derived from electrospray ionization Fourier transform ion cyclotron resonance mass spectrometry", *J. Agric. Food Chem.*, vol. 52, no. 17, pp. 5322–5328, August 2004.

[XU 10] XU Y., CHEUNG W., WINDER C.L. *et al.*, "VOC-based metabolic profiling for food spoilage detection with the application to detecting Salmonella typhimurium-contaminated pork", *Anal. Bioanal. Chem.*, vol. 397, no. 6, pp. 2439–2449, July 2010.

[YI 16] YI L. *et al.*, "Chemometric methods in data processing of mass spectrometry-based metabolomics: a review", *Anal. Chim. Acta*, vol. 914, pp. 17–34, 2016.

[ZAR 14] ZARDETTO S., BARBANTI D., ROSA M.D., "Formation of cholesterol oxidation products (COPs) and loss of cholesterol in fresh egg pasta as a function of thermal treatment processing", *Food Res. Int.*, vol. 62, pp. 177–182, August 2014.

[ZHA 10] ZHAO J., LIN H., CHEN Q. *et al.*, "Identification of egg's freshness using NIR and support vector data description", *J. Food Eng.*, vol. 98, no. 4, pp. 408–414, June 2010.

[ZHA 12] ZHANG H.-M. *et al.*, "Holistic quality evaluation of commercial white and red ginseng using a UPLC-QTOF-MS/MS-based metabolomics approach", *J. Pharm. Biomed. Anal.*, vol. 62, pp. 258–273, 2012.

[ZHA 14] ZHAO Y.Y., LIN R.-C., "Metabolomics in nephrotoxicity", in MAKOWSKI G.S. (ed.), *Advances in Clinical Chemistry*, vol. 65, Elsevier, 2014.

4

The Potential of Online Analysis Systems for the Chemotyping of the Egg Matrix

4.1. Introduction

The previous chapter focused on illustrating the potential of the latest analytical equipment to discover markers or groups of markers that are characteristic of LE spoilage. Depending on the resolving power of this chemotyping equipment, these markers would be either in the form of single molecules (in the case of high or low resolution spectrometers such as NMR or MS often coupled with chromatography), or in the form of groups of these same molecules, all of which are then identified unambiguously (NMR-HR and MS-HR, often coupled with chromatography), or in the form of a chemical signature reflecting the contribution of large chemical groups present in the sample (FTIR, NIRS, Raman).

Once identified, these markers could potentially be analyzed during LE production processes, using the same

---

Chapter written by Rémy COAT, Boris GOUILLEUX, Abdellah ARHALIASS, Jack LEGRAND, Gérald THOUAND, Estelle MARTINEAU, Ali ASSAF, Patrick GIRAUDEAU and Olivier GONÇALVES.

analytical approaches used for their characterization, but in technical forms suitable for online analysis. The quantitative results produced could be integrated into a process-type control loop (ultimately called "control-command"), which would allow, for example, on a production line, the modulation of its flows to prevent loss of lots (i.e. precocity markers indicating spoilage).

This chapter aims to deal with analytical approaches that could be used in process management as the latest generation of sensors and thus overcome a major challenge for the egg products industry, in particular by ensuring the quality of the products processed throughout the production line. We will focus first of all on the methods of analysis carried out directly during the processes, as this is one of the most effective ways to "monitor" LE spoilage. We will deal more particularly with the approaches exploiting micro-chromatographic and microspectrometric sensors such as those developed by SRA (with the interface coupled to the NeSSI process). So-called portable approaches, exploiting vibrational microspectrometry, will then be developed with Raman online analysis systems. Finally we will specifically address the very promising (so-called transportable) benchtop techniques stemming from very resolutive systems and usually reserved for large analytical service platforms. The new benchtop systems developed for routine and flow analyses will be presented, allowing the fine characterization of the samples in chromatography, NMR and mass spectrometry.

4.2. New analytical systems within processes

4.2.1. *Introduction*

The intensification of processes and their miniaturization are among the points listed for the factory of the future. This should, in the concept of the circular economy, introduce unitary operations that are eco-designed, intensified and modular. This involves focusing on process design, with technologies that will have to evolve towards major breakthroughs, thus making it possible to improve productivity. This transition to micro-processes will also require miniaturization of control techniques usually used for monitoring conventional processes. Thus in this dynamic, it becomes necessary to develop micro-analyzers and micro-sensors that can be placed within these miniaturized processes. It will no longer be a question of bringing "the sample to the analyzer" but "the analyzer to the sample". The current "online" or "offline" approaches will therefore evolve towards approaches that can be described as "in line", where experimentation times will be very small and the number of measurements multiplied. To do this, it is necessary to develop a new generation of analyzers that can monitor the quality of the product developed for a given unit operation while retaining, with regard to the reference methods, excellent robustness. This new generation of analyzers will have to be compact, coupled with micro-sampling systems in order to be directly integrated into the process and thus allow a closer analysis.

The need for a common standard for all industries to foster such development has emerged. The NeSSI standard or New Sampling/Sensor Initiative, project carried by the Center for Process Analysis and Control (http://cpac.apl.washington.edu/) or CPAC develops this concept. NeSSI is an "ad-hoc" unaffiliated group of more than 250 people, including user partners such as ExxonMobil, Chevron, Dow Chemical, Eastman Chemical, or instrumentalists such as ABB Analytical, Siemens, Fisher-Rosemount, Swagelok, Parker-Hannfin, CIRCOR etc.

4.2.2. *The NeSSI standard*

4.2.2.1. *Overview*

It was a patent filed in 1998 by Rob Dubois (engineer at Dow Chemical) that was behind the NeSSI initiative. The first miniature and modular sampling equipment was then proposed. This was an impetus driven by the CPAC which made it possible to introduce sampling and analysis improvements for manufacturing industries. In 2000, this concept called NeSSI was quickly supported by industry leaders including Dow Chemical (Rob Dubois) and Exxon Mobil (Peter Van Vuuren) [DUB 03] who aimed above all to develop a new standard based on this miniature and modular architecture (standard supported by ISA, or Instrumentation System and Automation). In 2002 the standard ANSI/ISA SP76.00.02-2002 was created [CAC 02].

4.2.2.2. *Technological objectives of NeSSI*

NeSSI technology proposes to change historical systems that are housed in the cabinets or shelters of sensors or complex chemical analyzers. These were coupled to the processes by tens or even hundreds of meters of pipes used to convey liquid or gaseous fluids to these cabinets. So instead of positioning itself "online", NeSSI is positioned "in line". Instead of proposing long transfer lines, NeSSI offers short transfer lines with reduced dead volumes, or even located within the process. Instead of offering a complex and specific design for each problem, NeSSI offers a simplified design, based on the use of modular blocks. Instead of proposing sample consumer analyses, NeSSI proposes to greatly reduce the quantities. Finally, instead of offering space-hungry technology, NeSSI offers micronized systems.

The NeSSI roadmap is based on a pyramidal development, where the technological base, essentially micromechanical elements, is already commercialized (generation 1). On this basis, communication elements are developed (local networks, mini-transducers, etc.; generation 2) to which a brand new generation of microanalyzers will be associated (generation 3 under development). All of these elements will be designed to ensure forward and backward compatibility.

The first generation of NeSSI elements, completed in 2002, is in line with the SP76 standard, that is, smaller dimensions (miniaturization), accompanied by the notion of modularity: the elements must be able to be associated like parts of “Lego”. It relates to surface mounted components that dispense fluids (liquids or gases) to analyzers and manual sample collection systems. Parker, Swagelok and Circor Tech already offer micro-modular and inter-exchangeable systems with, for example, flow control, pressure regulation or the possibility of filtering in the fluidic system. There are more than 60 components at the moment. These “plug and play” systems could eventually replace existing valves, filters, regulators, and current sensors (Figure 4.1).

Figure 4.1. *Example of a commercial assembly proposed by Swagelok including valves and digital or analog flow meters. Extract from Swagelok technical and commercial documentation (https://www.swagelok.com/ downloads/webcatalogs/ EN/MS-02-185.PDF)*

The second development phase of the NeSSI (generation 2) project is currently underway with the creation of new communication protocols between the sampling systems and networks of plants operating these modules. The main efforts are focused on the digitization of analogue signals recorded in the field and the means of transmitting them with reliable wireless technology (Wi-Fi). The development of a universal software interface for the management of these modules is also in the pipeline (Sensors Actuators Managers or SAM). The first prototype developed by Siemens Process Analytics was realized in 2006.

The third phase of NeSSI development concerns the development of adaptable microanalyzers on interfaces according to SP76 standards. This concerns approaches used in separation techniques such as micro-gas chromatography, liquid chromatography, steric exclusion, ionic, but also micro-spectrometric approaches that are vibrational (IR, NIR, Raman) or visible UV and fluorescence, mass spectrometry and even micro-NMR microanalyzers. Efforts are also being made to develop dielectric, impedance, conductivity micro-probes for measuring refractive indices, exploiting light scattering for determining particle size, measuring turbidity, pH, etc.

4.2.2.3. *Some examples of micro-analyzers compatible with the NeSSI standard*

GC microanalyzers

The Dutch company Concept to Volume (C2V) was one of the pioneers for the development of gas chromatographic micro-analyzers compatible with the NeSSI standard. It was bought by the giant Thermo Fisher Scientific in 2009 and since then Thermo has been marketing its “fast micro-GC” dedicated to gas analysis. This is interesting because the technology associated with micro-GC makes it able to work alone for laboratory or in-line use for analyses within processes. In addition, micronization increases the ramp performance at high temperatures (240 °C/min), decreases dead volumes (20nL), etc. Connectivity and multiplexing are also available, since standard interfaces such as USB or RS-232 are provided in series (Figure 4.2).

Figure 4.2. *Microchromatograph C2V marketed by Thermo Fisher. On the left are the microchromatography modules mounted in parallel, on the right one of the modules interfaced on a NeSSI system on a production line (extracted from Thermo Fisher technical and commercial documentation – DO1461, https://static.thermoscientific.com/images/D01461~.pdf). For a color version of this figure, see www.iste.co.uk/goncalves/ovoproducts.zip*

Current developments at CPAC

Many efforts have been made towards the development of fully automated systems dedicated to gas analysis (by micro-fast GC), the calibration of mixtures up to concentrations of ppb, using manometers in series and mass flow meters connected in parallel (Figure 4.3).

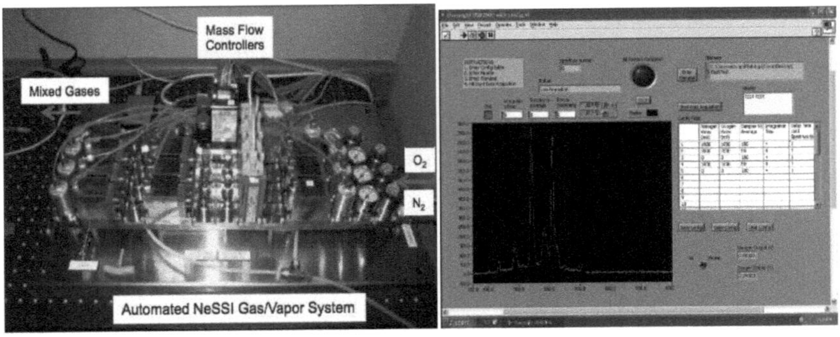

Figure 4.3. *Automatic gas mixer and its control system under Lab-view (excerpt from the presentation of Scott Gilbert and Brian Marquardt http://depts.washington.edu/cpac/NeSSI/40_IFPAC_09/Presentations/Marquardt%20-%20NeSSI.pdf – Brian Marquardt, Charles Branham, Kent Mann (University of Minnesota), Dave Veltkamp (C8 Medisys, Applied Physics Laboratory, Center for Process Analytical Chemistry, University of Washington, Seattle, WA 98105)). For a color version of this figure, see www.iste.co.uk/goncalves/ovoproducts.zip*

Projects are also being conducted on fiber optic sensors for the detection of gases (oxygen, ammonia, hydrogen, etc.) or certain organic compounds (simple solvents such as alcohols, esters or amines or certain chlorinated compounds, BTEX or even CO_2 and H_2S). It is in the context of vapochromic chemistry that these applications were developed because they allow an optical response specific to each analyte to be envisaged. The interest in this type of sensor is based on the simplicity of the design, the reversibility of measurements, the low power required, the low cost, the long-term stability, their high efficiency and sensitivity to a large number of analytes as well as wireless communication modes. Ultimately the plan is to develop oxygen sensors the size of a USB stick.

Liquid chromatography is also a topic under development at CPAC. This project is a key asset in the online control of LE spoilage, as it will broaden the control and monitoring of small molecules and metabolites currently analyzed primarily using offline approaches. This project is a major asset to the online control of LE spoilage, as it will develop process controls for the detection of small molecules and metabolites currently analyzed, mainly by offline approaches. Liquid phase chromatography for NeSSI systems uses a flux divider approach to sample the circulating matrix in process pipelines. Flow rates of around one microliter per minute are set. The sample is then sent to a microfluidic chip system where micro-mixing and micro-column separation steps follow. Detection is provided by a pulsed electrochemical system (Figure 4.4).

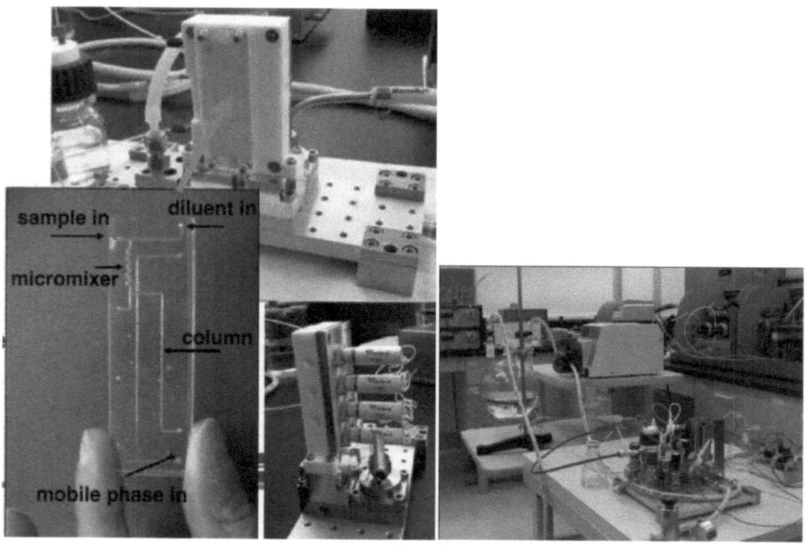

Figure 4.4. *Liquid microchromatography for NeSSI (excerpt from the presentation of Scott Gilbert and Brian Marquardt, http://depts.washington.edu/cpac/NeSSI/40_IFPAC_09/Presentations/Marquardt%20-%20NeSSI.pdf). On the left the micro-mixer and micro-fluidic chip system. On the right, a montage for the online control of a synthesis reaction in a balloon. For a color version of this figure, see www.iste.co.uk/goncalves/ovoproducts.zip*

NMR spectrometry is also envisaged for online measurements. Micro-NMR is a project involving miniaturized elements such as radio-frequency micro-coils with a size of about one mm (Figure 4.5). The picoSPIN 45

NMR microspectrometer (45 MHz very low magnetic field spectrometer, with a sampling volume of less than 20 µL) was proposed in 2012 and implements this concept. It is applied for mobile applications (NASA), control of pharmaceutical processes or specialty chemistry (Figure 4.6). The latest constitutes however a transition technology with a certain degree of maturity that still needs improvement for assessment of robust NMR signals for routine analysis (for that purpose see 4.4.2.3).

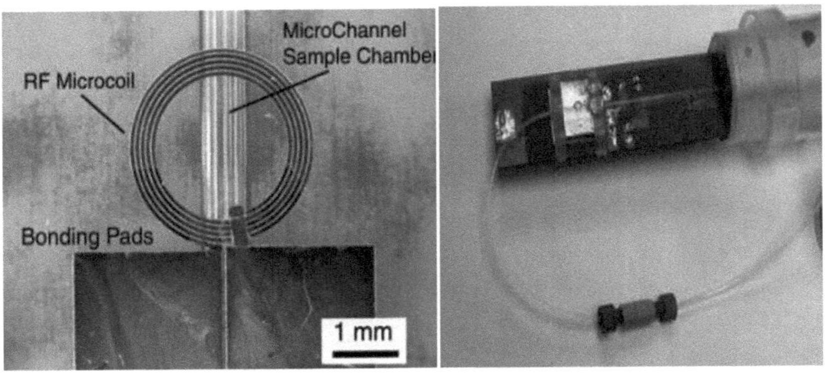

Figure 4.5. *Micro-NMR for NeSSI (excerpt from the presentation of Scott Gilbert and Brian Marquardt, http://depts.washington.edu/cpac/NeSSI/40_IFPAC_09/Presentations/Marquardt%20-%20NeSSI.pdf). For a color version of this figure, see www.iste.co.uk/goncalves/ovoproducts.zip*

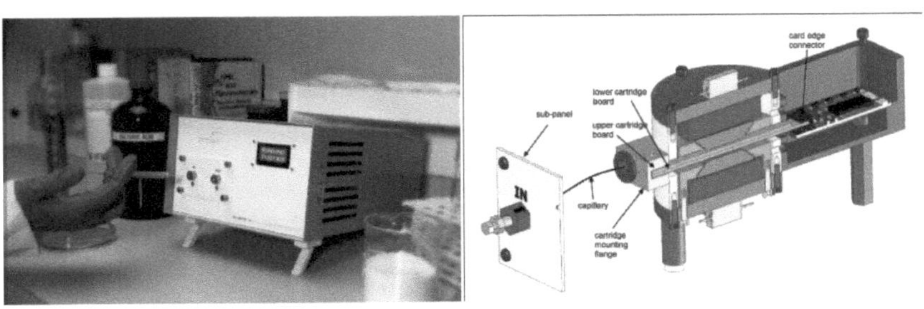

Figure 4.6. *NMR picoSPIN 45 micro-spectrometer (extract from technical documentation http://cpac.apl.washington.edu/files/picospincpacnessisher-man1105012.pdf). For a color version of this figure, see www.iste.co.uk/goncalves/ovoproducts.zip*

4.2.2.4. *Conclusion*

All these developments are either near completion (generation 1 NeSSI) or very promising, that is, generations 2 or 3, and thus allow applications on production lines of agri-food industries to be envisaged. It would therefore be possible to rapidly characterize the spoilage of egg products using probes with a broad analytical spectrum, providing information on the macroscopic quality of the samples (measurement of the gaseous sky, volatile compounds of interest, turbidity or pH), or by approaches targeting the chemical properties of these matrices (chemical profiles using vibrational spectroscopy, NMR, or research of marker metabolites in solution by in-line liquid micro-chromatography). However, this assumes that end-users will be ready to invest in these technologies, with implementations at several points in the production chains, which will in turn have to be optimized in terms of modularity. These profound changes may slow down the implementation of these new generations of sensors from laboratory approaches. Transient, more developed technologies requiring less effort can temporarily replace these approaches and will most likely converge towards NeSSI standards because they already take into account almost all of the technological objectives conveyed by the standard.

4.3. Transient analytical technologies, so-called transportable or bench-top

4.3.1. *Single-detection or mass spectrometry-based chromatographic analysis systems*

Portable micro-chromatography systems have been on the market for more than a decade. Their field of application has mainly focused on gas analysis since autonomous device developments were already possible without the need for significant technological breakthroughs. In Europe, a range of autonomous and efficient gas analyzers was proposed by the French company SRA. Several types of micro-GC are thus proposed for the analysis of gas mixtures, with analytical performances comparable to those of bench-top chromatographs, and often in a few minutes. These devices are modular, and conventionally consist of an automatic injection system, capillary separation column and micro-catharometric detection. The modular design makes it possible to use these devices in parallel, with their own carrier gas. Often very short response times are compatible with online analysis configurations. Moreover, with the portable and fully autonomous versions, these micro-GCs are true field analyzers that can be used in extreme environments (Figure 4.7).

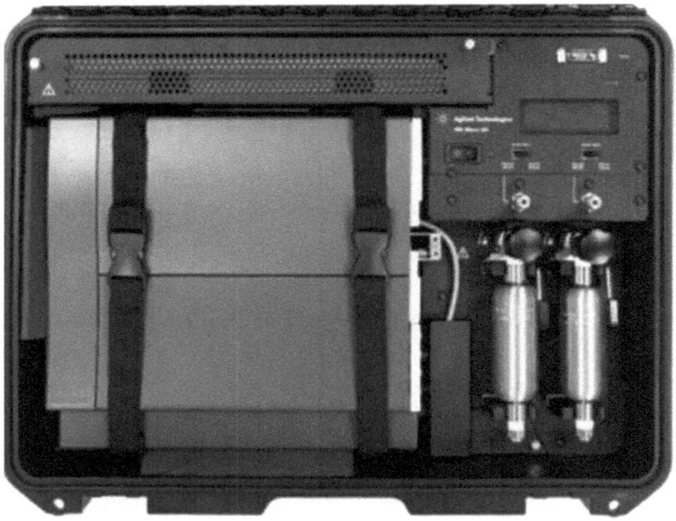

Figure 4.7. *Example configuration for a field micro-GC http://www.sra-instruments.com/fr/index.php?page=microgc-490.htm). On the left, the micro-CG, the upper right part showing the power supply and the lower right part the carrier gas reserves. For a color version of this figure, see www.iste.co.uk/goncalves/ ovoproducts.zip*

Indeed, even if some HPLC systems are already described as being autonomous and compact, such as Gilson's range of analytical and preparative chromatography, they are nonetheless dependent on mobile phase reservoirs, the volumes of which have only very slightly decreased. Finally, there is no true stand-alone HPLC system, even though the range proposed by ICON Scientific and redistributed by KNAUER in Europe offers a model the size of a shoebox and running on a battery (Figure 4.8).

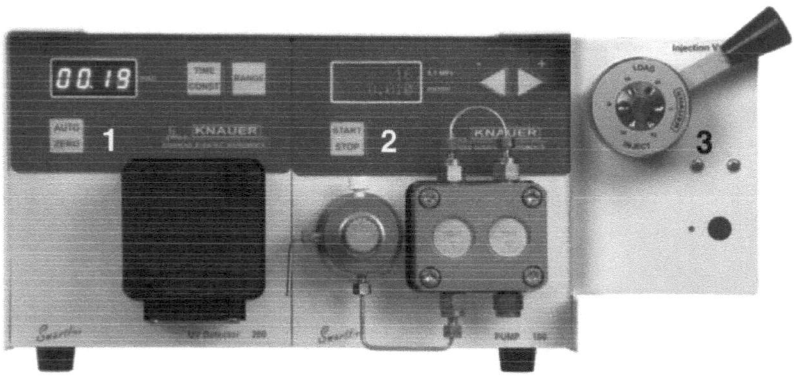

Figure 4.8. *Portable smartline HPLC. This is the smallest HPLC system currently marketed that offers analysis after manual injection, isocratic mode, with ultraviolet detection at a single wavelength (254 nm). Marketed by Knauer (Europe), Icon scientific (USA) (excerpt from Knauer technical documentation, http://www.knauer.net/fileadmin/user_upload/produkte/files/Dokumente/detectors/smartline/manuals/V7021-1_m_eg_dt_uv-detector_s200.pdf). 1: UV detector, 2: pump, 3: manual injection loop*

Truly autonomous chromatographic systems are therefore limited to the analysis of volatile compounds and mainly gases. The application of these approaches for detecting the spoilage of egg products is therefore limited to the gaseous atmosphere of these matrices and there are downfalls regarding their implementation. Nevertheless, recent developments in the coupling of mass detectors to micro-GC systems provide very encouraging short-term prospects. Indeed SRA currently offers a system combining micro-GC mass spectrometry analysis (quadrupole detector). Equipped with such a detection system, the performances of the micro-GC are increased in terms of sensitivity, since an increase of more than 10^3 can be measured according to targeted applications. The search for traces (such as volatile organic compounds or VOCs) thus becomes possible in production control thanks to pre-concentration solutions upstream of the analyses. The transportable solution can also operate in parallel thus increasing the fields of application (Figure 4.9).

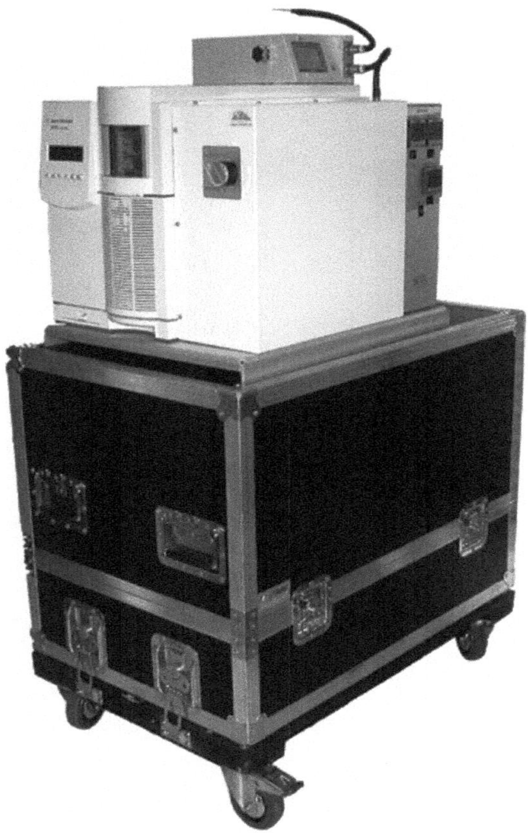

Figure 4.9. *Micro-GC coupled to a mass spectrometer as proposed in a transportable solution (extract from SRA commercial documentation, http://www.sra-instruments.com/fr/index.php?page=couplage.htm). For a color version of this figure, see www.iste.co.uk/goncalves/ovoproducts.zip*

Extensive investigations in less than one hour can be anticipated, with enumeration, identification and quantification of the chemical species present, that is with a level of analytical accuracy close to benchtop equipment (simple quadrupole GC-MS). With already commercialized solutions, little effort is required from a technical point of view to extend the analysis range to polar compounds. Adding a semi-automatic or automatic derivation system for the exhaustive analysis of non-volatile compounds from organic polar extracts would thus make it possible to obtain chemical impressions of an unaltered egg product matrix and to compare it with that of a spoiled matrix. Rapid control of the chemical status of the egg product could be done within a short time frame allowing the operators to intervene effectively in the production or distribution chain, i.e. in less than an hour (Figure 4.10).

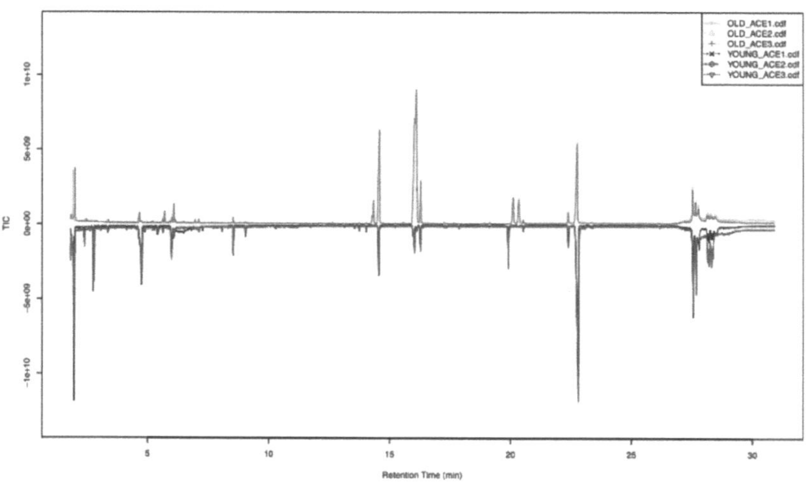

Figure 4.10. *Examples of chemical profiles of an aging LE sample compared to a fresh LE sample. The polar fractions analyzed here by GC-MS were obtained after LE deproteinization using acetone, fragmentation by polarity in a ternary mixture of methanol/chloroform/water and derivation with MSTFA according to [PHE 17]. Sky blue/green represent aging samples and purple/dark blue fresh samples. For a color version of this figure, see www.iste.co.uk/goncalves/ovoproducts.zip*

4.3.2. *Systems for obtaining chemical signatures using vibrational spectrometry*

Chemical signature approaches can also be obtained by less resolutive techniques than GC-MS in terms of information on the small molecules present in a sample, since they focus more on the nature of the chemical groups present in the studied matrix. Thus, instead of targeting signatures composed of differentiated chemical species, these approaches propose to characterize a sample on the basis of the diversity of the chemical groups composing it. Near-infrared and infra-red vibrational spectrometers have been cited in previous chapters and are approaches that can be routinely exploited for the control of spoilage in LEs. However, the limitations related to the presence of large amounts of water and proteins (FTIR) or the lack of robustness of the analytical technique (NIR) mean that applications are almost non-existent. Raman spectrometry is a particularly interesting candidate technique because of its potential for controlling the spoilage of food matrices. Indeed, there are many portable devices to consider for robust routine control in order to measure LE spoilage.

Recent technological advances in this field have made it possible to develop transportable spectrometers at moderate costs (between €20,000 and €60,000) thus offering powerful tools for the online monitoring of complex matrices [AVI 12]. It is now possible to observe the chemical structure of the samples in real time without preparation or extraction. It therefore becomes possible to multi-dimensionally map a sample as complex as an LE to identify the chemical compounds but also to precisely determine their spatial distribution [MEY 16]. It is the recent developments in microanalysis techniques that have considerably increased the analytical field of application of Raman spectrometry and make it possible to make this type of declaration. Indeed, current Raman systems have become so compact and robust that they are proposed by NASA and ESA for the development of space probes for the discovery of Martian soils [DIC 00, WYN 00]. In addition, the ability to perform remote analysis via optical fibers would facilitate control operations by offering manufacturers the ability to check their production remotely. The analysis of a sample in its different physical states is also now possible; solid compounds do not require special preparation (no pelletizing or thin film manufacturing), liquid samples or gas contained in transparent cells that can also be analyzed directly. At present, it is therefore possible to propose Raman spectrometry as an *in situ* or remote analysis tool in an industrial environment, for monitoring the chemical composition of LE, whether in single mode or multichannel mode. The Raman probe would be closer to the analyzed sample, in several places through multiplexing, and would be connected to the spectrometer by optical fibers thus allowing the transmission of signals over several hundred meters. In addition, Raman probes can currently operate at high pressure and high temperature making them compatible with most operations in agri-food processes and the egg processing industry in particular. The fact that they can operate when submerged (directly in the liquid whole egg) or through a sampling system thus respecting the sanitary conditions of the products (measured through a transparent glass window or in a reactor) limits the number of changes to the production lines for its implementation on site. Raman *in situ* could thus be exploited for the analysis of:

– solutions / reaction mixtures;

– emulsions;

– suspensions;

– gaseous skies in vials or reactors;

– raw materials;

– industrial effluents.

Figure 4.11 illustrates an example of a Raman probe used for the acquisition of a spectrum on a liquid whole egg product.

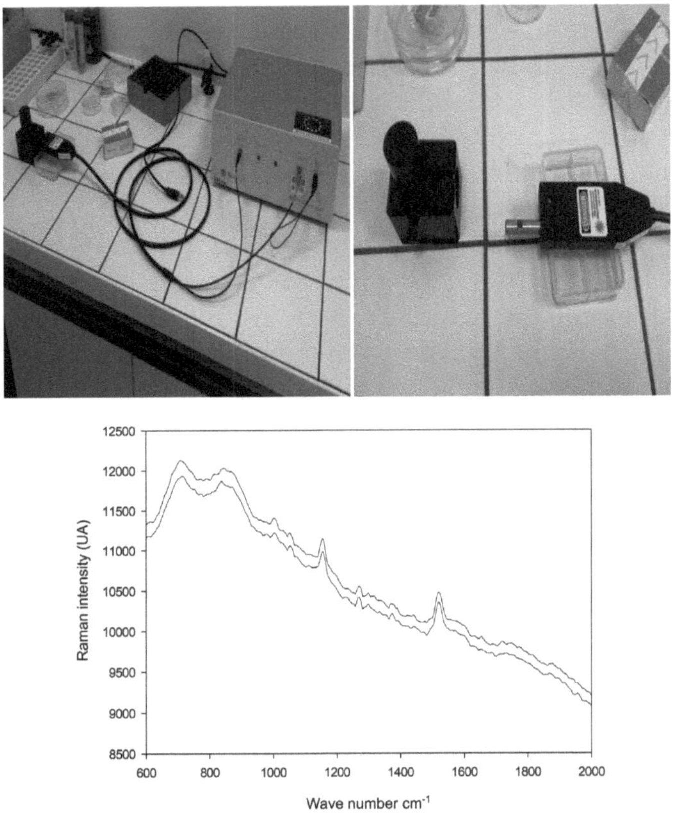

Figure 4.11. *Raman spectrum of a liquid whole egg product recorded at 532 nm using a compact Raman system (bayspec) (left) connected to a fiber optic probe (center). Superposition of two independent spectra recorded on an LE (right). The fluorescence effect of the LE sample distorts the baseline of the Raman spectrum recorded at 532 nm*

At the time of writing this section, there was only one example in the literature on the remote use of information from Raman spectrometry on the egg production line: it is in production control to determine the degree of contamination of egg white by egg yolk (thus limiting its foaming capacity even with low proportions > 0.01%). Cluff and colleagues in 2016 [CLU 16] demonstrated that there were significant differences in peak heights from spectra recorded on uncontaminated egg white compared to peak heights of spectra recorded on yolk-contaminated egg whites in the spectral region between 1080 and 1666 cm^{-1}. The method proved to be fast (from one to two minutes) and inexpensive as it was performed with entry-level benchtop Raman spectrometers (between €6,000 and €12,000). Figure 4.12 illustrates the results obtained during this study.

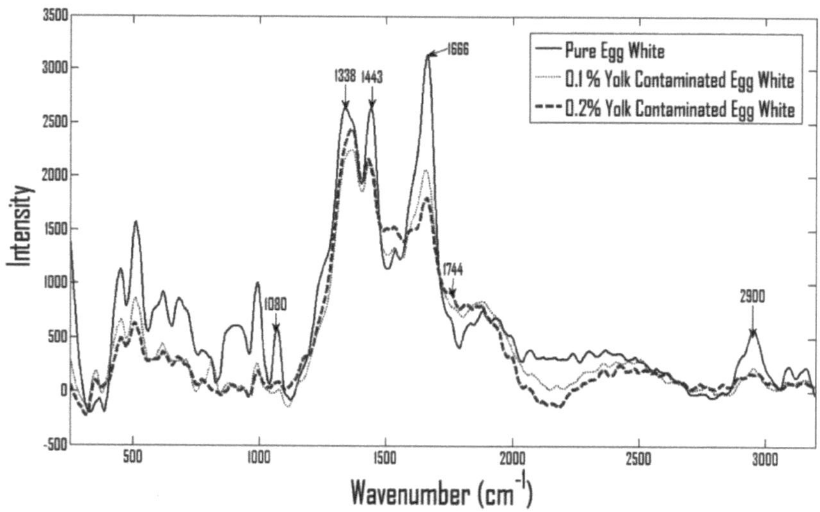

Figure 4.12. *Raman spectra of contaminated and pure eggs. Significant differences are indicated with the black arrows on the spectra [CLU 16]*

4.3.3. *Systems for obtaining chemical signatures by magnetic resonance spectrometry (portable NMR)*

High-field NMR (Chapter 3, section 3.4.3) has great potential for the analysis of complex mixtures such as egg products. However, its accessibility remains limited by the cost and bulk of high-field spectrometers, but also by maintenance costs and the associated high degree of technicality. High-field NMR is thus mainly accessible to small and medium-sized companies through links with academic platforms or large companies. Low-field NMR is much more accessible, but has long been limited to relaxometry measurements described in section 3.4.1.2.2 in Chapter 3. The main reason is that the permanent magnets used in the relaxometers do not have sufficient homogeneity to allow the acquisition of high resolution spectra. However, the 2010s saw the emergence of a new generation of low magnetic field NMR spectrometers, based on a new generation of permanent magnets (Hallbach magnets). These magnets make it possible to generate a magnetic field of around 1 Tesla, sufficiently homogeneous for the acquisition of high resolution spectra [KUS 11]. Several manufacturers offer this equipment, which stands out by being transportable (a few tens of kg) and compact enough to be placed on a bench or under a hood. The purchase cost is 10 to 20 times lower than that of a high-field spectrometer, and its operational cost is almost negligible. Indeed, the use of a permanent magnet does not require cryogenic fluids, unlike high-field spectrometers that use superconducting magnets. The operating cost of these low-field spectrometers is between 1 and 2 euros per hour of use, making it an attractive tool for many applications. NMR spectroscopy is no longer limited to large laboratories or large companies, and has become accessible at a reasonable cost.

As in high-field liquid NMR, the analysis is carried out in solution in a 5 mm diameter tube, for a total sample volume of about 700 µL. However, unlike high-field analysis, the use of a deuterated solvent is not essential because low-field spectrometers are equipped with an internal field-frequency lock system. This feature simplifies sample preparation and reduces the cost of the analysis, but it is then necessary to implement approaches to selectively suppress the solvent signal to detect peaks of interest [GOU 17]. On the other hand, most low-field NMR devices are equipped with a continuous flow detection system that is particularly suitable for online monitoring of chemical or biochemical processes [DAL 12]. The "online" character of this equipment makes it particularly attractive for industrial applications, but also for the study of reaction mechanisms in chemistry.

Obviously, this miniaturization has a cost in terms of resolution and sensitivity. Regarding the sensitivity, concentrations of around a few millimoles per liter can be detected in a few minutes. Low-field NMR spectroscopy is therefore limited to the study of relatively concentrated samples. On the other hand, due to a reduced spectral range, the resolution is limited, characterized by a strong overlap of peaks. Multidimensional acquisition methods can be used to improve the dispersion [GOU 16], but in general, it is difficult to imagine using low-field NMR spectroscopy to carry out the structural identification or the targeted quantification of compounds in complex mixtures. Still, low-field NMR can be applied to complex samples to obtain a characteristic profile consisting of a sample fingerprint.

At the time of writing this chapter, low-field NMR had never been applied to the study of egg products, but it must be noted that this is a very recent technology. In the agri-food field, low-field NMR spectroscopy has recently been applied to the detection of olive oil adulteration processes, associating a ^{1}H-NMR low-field acquisition with a statistical processing of data [PAR 14]. In this study, the performance of low-field NMR was compared to Fourier Transform Infrared Spectroscopy (FTIR), an analytical approach comparable to low-field NMR in terms of cost and accessibility. The authors demonstrated a better specificity of NMR for a sensitivity comparable to FTIR, thus demonstrating the potential of low-field NMR spectroscopy as a profiling tool.

Figure 4.13 illustrates the potential of low-field NMR on samples of egg products analyzed on a low magnetic field spectrometer (1 Tesla, Figure 4.13a).

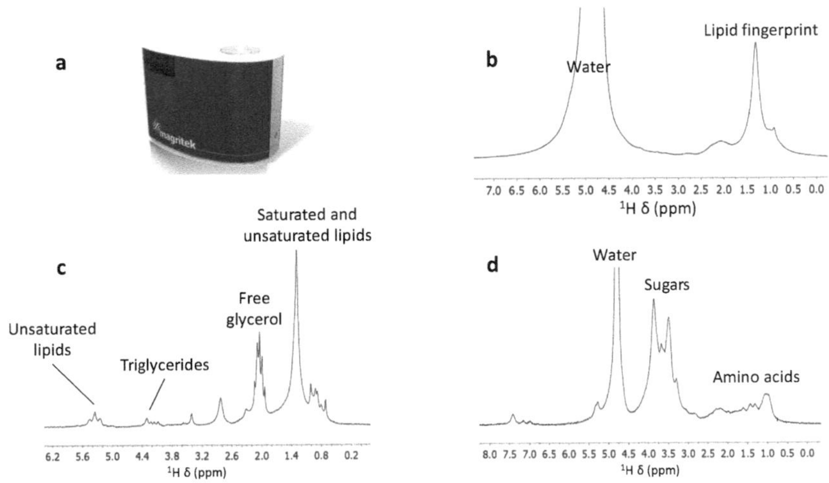

Figure 4.13. *Illustration of the potential of liquid-phase low-field NMR spectroscopy for the analysis of egg products. (a) Example of a low-field NMR spectrometer (1 T Spinsolve model built by Magritek). (b) Proton NMR spectrum recorded in five minutes obtained at 1 Ton a lyophilized whole egg sample and reconstituted in an aqueous phase for NMR analysis. (c) Spectrum of the apolar fraction (lyophilized and then reconstituted in deuterated acetone) recorded in four minutes. (d) Spectrum of the polar fraction (lyophilized and then reconstituted in D_2O) recorded in four minutes. Experiments carried out in the Chemistry and Interdisciplinarity Laboratory: Synthesis, Analysis, Modeling, at the University of Nantes*

The spectrum of a whole egg sample (Figure 4.13b) is relatively uninformative, but it can distinguish between the lipid signal and the water signal. The spectra of the apolar (Figure 4.13c) and polar (Figure 4.13d) fractions show a lot of exploitable information. On the apolar fraction spectrum, the CH_2 lipid groups (saturated and unsaturated) give rise to the most intense peak at 1.2 ppm, whereas the contribution of the unsaturated lipids can be separated thanks to the peak at 5.4 ppm. The peak at 2.1 ppm corresponds to the free glycerol signal, whereas the glycerol chains present in the triglycerides appear at around 4.2 ppm. On the polar fraction spectrum, the contribution of glucose can be separated from that of amino acids, without the metabolites being individually identified as on the high-field spectrum.

These preliminary tests therefore suggest that low-field NMR, a very recent technology, has great potential for the acquisition of chemical fingerprints of egg products, in particular from the lipid or metabolic extracts derived from egg products. Combined with statistical processing of data, low-field NMR spectroscopy could be an excellent alternative to existing portable techniques such as FTIR, Raman or mass spectrometry-based separation approaches. To control the chemical state of LEs on a production line, this approach also has several advantages with regard to chromatography profiling techniques in particular. Indeed, without the need for special sample processing, it is possible to obtain a robust, semi-quantitative or even quantitative chemical image of a healthy LE sample or one undergoing spoilage (relatively low investment costs: a spectrometer costs less than €80,000 for a mid-range version). Moreover, the ability to automate the processing of information in a real command control loop makes it a technology with great application potential in the agri-food and egg product industries [GUT 16, MEY 16].

4.4. Conclusion

Alongside the development of very high resolution analytical systems, which seek to take advantage of the main features of NMR in terms of versatility and mass spectrometry in terms of sensitivity, a whole range of transportable devices derived from these technologies is developing. NMR, FTIRs or Raman spectrometers, or chromatographs often coupled with mass spectrometers, are now available in affordable, transportable, robust and quasi-compact formats. Some developments remain for the specific processing unique to each sample for offline or online measurements, but the efforts required are minimal. This suggests a rapid development of the use of these approaches for the routine control of the chemical quality of LE matrices or more generally agri-food matrices. Efforts to develop measurement systems as part of the process, or "in-line", are the most revolutionary at the moment. Indeed, the NeSSI initiative led by the Center for Process Analysis and Control should benefit from a whole new range of sensors derived from the complex analytical techniques usually reserved for laboratories and analytical chemistry platforms. This new generation of sensors will be miniaturized, thus respecting the requirements for future factories for the development of intensified, compact and modular processes. It will also perform better than current techniques while gaining simplicity of use and robustness.

4.5. Bibliography

[AVI 12] AVILA T.C., POPPI R.J., LUNARDI I. *et al.*, "Raman spectroscopy and chemometrics for on-line control of glucose fermentation by Saccharomyces cerevisiae", *Biotechnol. Prog.*, vol. 28, no. 6, pp. 1598–1604, 2012.

[CAC 02] COMPOSITIONAL ANALYZERS COMMITTEE (ISA), ISA-76.00.02-2002 Modular Component Interfaces for Surface-Mount Fluid Distribution Components – Part 1: Elastomeric Seals, Instrumentation, Systems, and Automation Society (ISA), 2002.

[CLU 16] CLUFF K., KONDA NAGANATHAN G., JONNALAGADA D. *et al.*, "Determination of yolk contamination in liquid egg white using Raman spectroscopy", *Poult. Sci.*, vol. 95, no. 7, pp. 1702–1708, 2016.

[DAL 12] DALITZ F., CUDAJ M., MAIWALD M. *et al.*, "Process and reaction monitoring by low-field NMR spectroscopy", *Prog Nucl Magn Reson Spectrosc*, vol. 60, pp. 52–70, 2012.

[DIC 00] DICKENSHEETS D.L., WYNN-WILLIAMS D.D., EDWARDS H.G. *et al.*, "A novel miniature confocal microscope/Raman spectrometer system for biomolecular analysis on future Mars missions after Antarctic trials", *J. Raman Spectrosc.*, vol. 31, no. 7, pp. 633–635, 2000.

[DUB 03] DUBOIS R., VAN VUUREN P., GUNNELL J.J., "NeSSI™ (New Sampling/Sensor Initiative) Generation II Specification", *Cent. Process Anal. Chem. CPAC Univ. Wash. Seattle WA*, pp. 122–130, 2003.

[GOU 16] GOUILLEUX B., CHARRIER B., AKOKA S. *et al.*, "Ultrafast 2D NMR on a benchtop spectrometer: Applications and perspectives", *Trends. Anal. Chem.*, vol. 83, pp. 65–75, 2016.

[GOU 17] GOUILLEUX B., CHARRIER B., AKOKA S. *et al.*, "Gradient-based solvent suppression methods on a benchtop spectrometer", *Magn. Reson. Chem.*, vol. 55, pp. 91–98, 2017.

[GUT 16] GUTHAUSEN G., "Analysis of food and emulsions", *TrAC Trends Anal. Chem.*, vol. 83, pp. 103–106, 2016.

[KUS 11] KUSTER S.K., DANIELI E., BLUMICH B. *et al.*, "High-resolution NMR spectroscopy under the fume hood", *Phys. Chem. Chem. Phys.*, vol. 13, no. 29, pp. 13172–13176, 2011.

[MEY 16] MEYER K., KERN S., ZIENTEK N. *et al.*, "Process control with compact NMR", *TrAC Trends Anal. Chem.*, vol. 83, pp. 39–52, 2016.

[PAR 14] PARKER T., LIMER E., WATSON A.D. *et al.*, "60 MHz 1H NMR spectroscopy for the analysis of edible oils", *Trends Anal Chem*, vol. 57, no. 147–158, 2014.

[PHE 17] PHELIPPÉ M., "Characterization of an easy-to-use method for the routine analysis of the central metabolism using an affordable low-resolution GC-MS system: Application to Arthrospira platensis", *Analytical and Bioanalytical Chemistry*, November 2017.

[WYN 00] WYNN-WILLIAMS D., EDWARDS H., "Proximal analysis of regolith habitats and protective biomolecules in situ by laser Raman spectroscopy: overview of terrestrial Antarctic habitats and Mars analogs", *Icarus*, vol. 144, no. 2, pp. 486–503, 2000.

Conclusion

The biological complexity of egg products makes them a matrix of great chemical diversity with a compartmentalization that does not favor the systematic detection of small molecules, which, however, reflects the degree of freshness or spoilage of the egg product during its manufacturing process. Since these metabolites are highly diverse in terms of physico-chemical properties, they can only be analyzed through laboratory analytical techniques. Thus approaches resulting from metabolomics all have the potential to detect these small compounds, but this requires time-consuming sample-preparation steps that are often incompatible with short response times, which are sought preferentially when carrying out such processes. Moreover, access to these analytical techniques is limited to service platforms, which prevents their routine use in production environments.

The latest instrumental developments, making it possible for the general public to access these highly resolutive analytical techniques, suggest completely different chemical control methods. Indeed, portable techniques are increase-ingly proposed as substitutes for conventional benchtop devices. Thus, mobility, robustness and ease of use are ultimately associated with techniques that until now have been reserved for chemical engineer-doctors. This is particularly the case for low magnetic field NMR systems whose analytical capabilities seem to be able to chemically profile samples extracted from egg products.

An important technological breakthrough is currently underway with the efforts made on the NeSSI standard. This time, transportable high-resolution devices are no longer proposed, but instead a whole new generation of sensors, directly inspired by chromatographic systems or nuclear magnetic resonance. The main principles of these separation techniques, whether they are in the gas or liquid phase, or non-invasive spectrometric techniques such as NMR, are adapted and miniaturized. Thus, it is no longer online analysis techniques that are emerging, but in-line analysis techniques that are being or will soon be proposed. The develop-ments concerning liquid chromatography (based on micro or nano-fluidic systems) or NMR (with micro-magnets) are the most promising because they make it possible to extend the range of possibilities to thermosensitive and water-soluble markers. Thus, the metabolites identified as being representative of biotic or abiotic spoilage of egg product matrices could in the near future be analyzed quickly, thus proposing early detection systems for the spoilage of these complex matrices.

List of Authors

Marc ANTON
INRA Angers-Nantes
Pays de la Loire
Nantes
France

Abdellah ARHALIASS
GEPEA – UMR CNRS 6144
CRTT
Saint-Nazaire
France

Ali ASSAF
GEPEA – UMR CNRS 6144
CRTT
Saint-Nazaire
France

Florence BARON
STLO – UMR INRA 1253
AGROCAMPUS OUEST
(P3AN)
Rennes
France

Rémy COAT
GEPEA – UMR CNRS 6144
CRTT
Saint-Nazaire
France

Frédérique COURANT
Hydrosciences
UMR CNRS 5569
University of Montpellier 1
France

Patrick GIRAUDEAU
CEISAM – UMR CNRS 6230
Nantes
France

Olivier GONÇALVES
GEPEA – UMR CNRS 6144
CRTT
Saint-Nazaire
France

Boris GOUILLEUX
CEISAM – UMR CNRS 6230
Nantes
France

Sophie JAN
STLO – UMR INRA 1253
AGROCAMPUS OUEST
(P3AN)
Rennes
France

Valérie LECHEVALIER
STLO – UMR INRA 1253
AGROCAMPUS OUEST
(P3AN)
Rennes
France

Jack LEGRAND
GEPEA – UMR CNRS 6144
CRTT
Saint-Nazaire
France

Estelle MARTINEAU
CEISAM – UMR CNRS 6230
Nantes
France

Anne MEYNIER
INRA Angers-Nantes
Pays de la Loire
Nantes
France

Françoise NAU
STLO – UMR INRA 1253
AGROCAMPUS OUEST
(P3AN)
Rennes
France

Anne THIERRY
STLO – UMR INRA 1253
AGROCAMPUS OUEST
(P3AN)
Rennes
France

Gérald THOUAND
GEPEA – UMR CNRS 6144
CRTT
Saint-Nazaire
France

Index

I, M

N, P

Q, S

FSC
www.fsc.org
MIX
Papier | Fördert
gute Waldnutzung
FSC® C083411

Kolibri 360 GmbH
Ferdinand-Jühlke-Straße 7
99095 Erfurt, Deutschland
produktsicherheit@kolibri360.de